AF757642

Foods, Nutrients and Food Ingredients with Authorised EU Health Claims

Related Titles

Foods, Nutrients and Food Ingredients with Authorised EU Health Claims, Volume 1
(978-0-85709-842-9)

Foods, Nutrients and Food Ingredients with Authorised EU Health Claims, Volume 2
(978-1-78242-382-9)

Global Legislation for Food Contact Materials
(978-1-78242-014-9)

Advances in Food and Beverage Labeling: Information and Regulations
(978-1-78242-085-9)

Woodhead Publishing Series in Food Science, Technology and Nutrition

Foods, Nutrients and Food Ingredients with Authorised EU Health Claims

Volume 3

Edited by

Michele J. Sadler

ELSEVIER

WP
WOODHEAD PUBLISHING
An imprint of Elsevier

Woodhead Publishing is an imprint of Elsevier
The Officers' Mess Business Centre, Royston Road, Duxford, CB22 4QH, United Kingdom
50 Hampshire Street, 5th Floor, Cambridge, MA 02139, United States
The Boulevard, Langford Lane, Kidlington, OX5 1GB, United Kingdom

Copyright © 2018 Elsevier Ltd. All rights reserved.

No part of this publication may be reproduced or transmitted in any form or by any means, electronic or mechanical, including photocopying, recording, or any information storage and retrieval system, without permission in writing from the publisher. Details on how to seek permission, further information about the Publisher's permissions policies and our arrangements with organizations such as the Copyright Clearance Center and the Copyright Licensing Agency, can be found at our website: www.elsevier.com/permissions.

This book and the individual contributions contained in it are protected under copyright by the Publisher (other than as may be noted herein).

Notices
Knowledge and best practice in this field are constantly changing. As new research and experience broaden our understanding, changes in research methods, professional practices, or medical treatment may become necessary.

Practitioners and researchers must always rely on their own experience and knowledge in evaluating and using any information, methods, compounds, or experiments described herein. In using such information or methods they should be mindful of their own safety and the safety of others, including parties for whom they have a professional responsibility.

To the fullest extent of the law, neither the Publisher nor the authors, contributors, or editors, assume any liability for any injury and/or damage to persons or property as a matter of products liability, negligence or otherwise, or from any use or operation of any methods, products, instructions, or ideas contained in the material herein.

Library of Congress Cataloging-in-Publication Data
A catalog record for this book is available from the Library of Congress

British Library Cataloguing-in-Publication Data
A catalogue record for this book is available from the British Library

ISBN: 978-0-08-100922-2 (print)
ISBN: 978-0-08-101006-8 (online)

For information on all Woodhead publications visit our website at
https://www.elsevier.com/books-and-journals

ELSEVIER
Book Aid International
Working together to grow libraries in developing countries
www.elsevier.com • www.bookaid.org

Publisher: André Gerhard Wolff
Acquisition Editor: Nancy Maragioglio
Editorial Project Manager: Karen Miller
Production Project Manager: Omer Mukthar
Designer: Mark Rogers

Typeset by Thomson Digital

Volume 3 of Foods, Nutrients and Food Ingredients With Authorised EU Health Claims *is dedicated to Dr Janice Harland RNutr FAfN (1954–2017). Following a successful career in the food industry, Janice set up her own company (Harland Hall Associates, United Kingdom) and worked as a nutrition consultant, lecturer and author. Janice was an enthusiastic contributor to all of the three Volumes of this series, having contributed chapters on barley and oat β-glucans and on children's claims in Volume 1, on oat and barley grain fibre and on wheat bran fibre in Volume 2 and on sugar beet fibre is this current Volume. This reflects Janice's specialism in nutrition and health claim legislation, the breadth of her knowledge and experience in the area and her broader aim of wanting to disseminate evidence-based nutrition and health science to a wider audience. Colleagues have reflected that Janice will be missed for her professionalism, scientific acumen, wealth of nutrition knowledge, support and advice which she was always willing to give, and for the contribution she had still to make to nutrition science.*

Contents

† Deceased

List of contributors

Ursula Arens Freelance nutrition writer, London, United Kingdom

Heike A. Bischoff-Ferrari University Hospital and University of Zurich; University Clinic for Acute Geriatric Care, City Hospital Waid, Zurich, Switzerland

Eric Chappuis Naturalpha, Loos, France

Jess Funnell Bournemouth University, Poole, United Kingdom

Philippa S. Gibson University of Surrey, Guildford, United Kingdom

Klaus G. Grunert MAPP Centre, Aarhus University, Aarhus C, Denmark

Janice I. Harland[†] Harland Hall Associates, Poulton, Cirencester, United Kingdom

Sophie Hieke European Food Information Council, Brussels, Belgium

Nicole J. Kellow Monash University, Melbourne, VIC, Australia

Sarah Kuczora Dudley Metropolitan Borough Council, Dudley, United Kingdom

Susan A. Lanham-New University of Surrey, Guildford, United Kingdom

Katherine M. Livingstone Institute for Physical Activity and Nutrition, Deakin University, Geelong, VIC, Australia

Michele J. Sadler Rank Nutrition Ltd, Bethersden, Kent, United Kingdom

Stephan Theis BENEO Institute, Obrigheim, Germany

Karen Z. Walker Monash University, Melbourne, VIC, Australia

[†] Deceased

Preface

Health claims are a common tool for promoting food and drink products and for communicating their health attributes to consumers. All health claims used in Europe must be preapproved with the aim of protecting consumers from misleading or false claims, and as this Volume goes to print, there are currently 267 approved claims in Europe. Following the success of Volumes 1 and 2 of *Foods, Nutrients and Food Ingredients With Authorised EU Health Claims*, the current Volume (Volume 3) aims to compile relevant scientific and technical information relating to health claims authorised in the European Union that have not so far been covered in the previous two Volumes.

Claims within the three regulatory categories are covered, i.e.:

1. Article 13.1 claims, referring to:
   a. The role of a nutrient or other substance in growth, development and the functions of the body.
   b. Psychological and behavioural functions.
   c. Slimming, weight control, a reduction in the sense of hunger, an increase in the sense of satiety or the reduction of the available energy from the diet and based on generally accepted scientific evidence.
2. Article 13.5 claims, based on newly developed scientific evidence and/or including a request for the protection of proprietary data.
3. Article 14 claims, specifically
   a. Article 14.1(a) claims referring to the reduction of disease risk.
   b. Article 14.1(b) claims referring to children's development and health.

Additionally, Volume 3 aims to inform the reader about the latest regulatory developments relating to health claims in Europe, continuing the updates provided in Volumes 1 and 2. Also included in Volume 3 is a chapter dedicated to the consumer understanding of claims. Though consumer understanding is a key aim of the European health claim legislation, many of the authorised claim wordings are expressed in highly scientific language that is unlikely to be familiar to many consumers. Hence this is a key issue that is afforded particular consideration in the current Volume.

Continuing with the same format as for Volumes 1 and 2, each chapter focuses on foods, nutrients or food ingredients that have authorised health claims in Europe. Where available, the information aims to cover:

- characterisation of the food, nutrient or food ingredient;
- details of the health claim;
- a summary of the substantiating and supporting evidence for the claims;
- authorised wording and how the health claims benefit consumers;
- conditions of use, warning labels and any safety issues;
- how the food, nutrient or food ingredient can be incorporated into a balanced diet;

- details of available research on consumer understanding of health claim wording and use of flexibility of the wording in practice;
- importance or potential of authorised health claims in product development;
- how authorised claims are being used in practice, on products and in commercial communications such as websites and advertising in different EU Member States and
- other regulatory issues or legislation relevant to use of the claim, e.g. novel foods and medicines regulations.

In some cases, where the claims are newly authorised, their use on products and in marketing has not yet become established, making it premature to provide a detailed discussion of how the claims are being used in practice and examples of flexibility of the wording. Relevant information about consumer understanding of the claims is also not uniformly available for all of the claims. The content of the Volume does not address general, non-specific health benefits, which are permitted, provided they are accompanied by a specific authorised health claim. It is not the purpose of the book to address nutrition claims or related non-authorised health claims, though in some cases their discussion is relevant to the main content of the chapter.

Volume 3 thus continues the aims of Volumes 1 and 2 by bringing together technical, scientific and regulatory information about foods, nutrients and food ingredients with authorised EU health claims into one resource, and providing a unique compilation of subjects and regulatory issues pertaining to the development of food products with a health proposition.

For ease of accessibility the chapters are organised as follows:

Part I: Regulatory background
Part II: Authorised reduction of disease risk claims
Part III: Health claims based on emerging science and proprietary claims
Part IV: General function claims

With the addition of Volume 3 to the series, the majority of authorised health claims are now covered. However, as the most recent claims were authorised too late for inclusion, these are briefly referred to in Chapter 1.

The envisaged readership includes health professionals, food business operators, technical roles within the food industry, scientists and academics, students of nutrition and food sciences, law enforcement officers, regulators and health professionals. The aims are to provide a resource covering relevant background information about foods, nutrients and ingredients with authorised EU health claims and the scientific support for such claims. The intended purpose of the book is to enable readers with roles in the food industry to access background information that is relevant to the development of products with a health proposition and to highlight the range of potential issues that may be relevant in gaining approval and ultimately enabling use of both new and existing claims. A further aim is to enable R&D managers, scientists and academics to benefit from the experience gained so far concerning the type, quality and quantity of scientific evidence that is needed for successful health claim applications, to help with planning both scientific and consumer research.

The EU register of nutrition and health claims (available at http://ec.europa.eu/food/safety/labelling_nutrition/claims/register/public/?event=register.home) is updated at regular intervals as decisions are made on new applications. As the list is subject to change, the onus is on the reader always to consult the EU register for the most up-to-date information and to obtain the officially approved wording for health claims and the approved conditions of use. Likewise, the reader should also consult the relevant European Food Safety Authority (EFSA) opinions when considering the use of health claims, as this book is intended to provide a broad-brush coverage of the main issues and is not a legal text.

A multitude of guidance has been developed to assist with compliance with Regulation 1924/2006 and the reader is advised to consult these documents also. A number of trade bodies have also developed guidance specific to their sectors. However, not all requirements of the Regulation have yet been clarified by the relevant authorities, and a number of issues of interpretation of the Regulation continue to remain unclear. These and the many related issues are discussed in Chapter 1 which provides an update on regulatory developments, further to that covered in Volumes 1 and 2.

Once again, my sincere thanks are due to the authors who have so expertly and diligently contributed to Volume 3 within the required deadlines, and to the support team at Woodhead Publishing. Their valuable support throughout the whole process has once again enabled timely publication of the Volume.

Michele J. Sadler

Part One

Regulatory Background

European health claims: regulatory developments

1

Michele J. Sadler
Rank Nutrition Ltd, Bethersden, Kent, United Kingdom

1.1 Introduction

This chapter discusses developments since January 2015 relating to the regulation of health claims in Europe, which is governed by the Nutrition and Health Claims Regulation (EU, 2006a). The equivalent chapter in Volume 2 (Sadler, 2015) covers regulatory developments from September 2013, while developments prior to that are covered in Volume 1 (Binns, 2014).

Recent developments of note are a slower rate of new health claim applications through the available application routes (EU, 2006a):

- Article 13.5: claims based on newly developed scientific evidence and/or requesting the protection of proprietary data.
- Article 14.1(a): reduction of disease risk claims.
- Article 14.1(b): children's development and health claims.

Of the candidate 'general function' claims previously submitted by Member States (prior to 31 January 2008) under Article 13.1 (Binns, 2014), over 2000 claims for 'botanicals' remain on-hold, pending a wide-ranging regulatory review. The requirement in the Nutrition and Health Claims Regulation (EU, 2006a) to set nutrient profiles, such that foods not meeting certain nutritional criteria would not be able to carry health claims, is also under review. Though rules have been published to apply for generic descriptors, at the time of writing, none have as yet been approved. These issues are discussed along with other general trends, reference to new guidance documents, and possible effects of the changing political landscape, particularly in the United Kingdom in light of the decision to leave the European Union (so-called Brexit).

1.2 Health claim applications

The slower pace of new health claim applications reflects the associated difficulty, the uncertainty and the high costs of both undertaking new studies and compiling applications which have resulted in less commercial interest in winning health claim approvals in Europe. Of applications that have been made, negative EFSA opinions continue to outweigh positive ones, and the time taken to authorise claims remains variable, depending on particular issues that require discussion. The numbers of authorised and non-authorised health claims by category as of July 2017 are shown in Table 1.1.

Foods, Nutrients and Food Ingredients with Authorised EU Health Claims. http://dx.doi.org/10.1016/B978-0-08-100922-2.00001-2
Copyright © 2018 Elsevier Ltd. All rights reserved.

Table 1.1 Number of authorised and non-authorised health claims by category (as of July 2017)

Category (article)	Authorised	Non-authorised
13.1 'General function claims'	229	1875
13.5 Claims based on newly developed scientific evidence	6	108
13.5 Claims based on proprietary data	5	
14.1(a) Reduction of disease risk claims	14	24
14.1(b) Children's development and health claims	12	44
Number of 13.5 and 14.1(a) and (b) health claims under discussion following receipt of the EFSA opinion is ~50		

1.2.1 Article 13.5 claims

1.2.1.1 Authorisations based on newly developed scientific data

Two claims were authorised in the first half of 2017, too late for inclusion in Volume 3. These were for creatine in conjunction with resistance training and 'improvement in muscle strength' in adults aged over 55 years, and for lactitol and 'maintenance of normal defecation'. Lactitol has since received novel foods approval for nutritional or physiological purposes, for use in capsule or tablet forms of food supplements intended for adults at a maximum daily dose of 20 g.

Four Article 13.5 applications that are also not the subject of new chapters were for claims that had already been assessed with a favourable outcome under Article 13.1. They related to two claims: one for non-fermentable carbohydrates and 'consumption of foods/drinks containing [name of all used non-fermentable carbohydrates] instead of fermentable carbohydrates contributes to the maintenance of tooth mineralisation', and the second for non-digestible carbohydrates and 'consumption of foods/drinks containing [name of all used non-digestible carbohydrates] instead of sugars induces a lower blood glucose rise after their consumption compared to sugar-containing foods/drinks'. These have been authorised as Article 13.5 claims, and the related Article 13.1 claims were discussed in Volume 1 (Shortt, 2014).

Authorisation of the claim for glycaemic carbohydrates and 'recovery of normal muscle function after highly intensive and/or long-lasting physical exercise leading to muscle fatigue and the depletion of glycogen stores in skeletal muscle' and the claim for sugar beet fibre are both discussed in this volume (Chappuis, 2017; Harland, 2017).

1.2.1.2 Authorised proprietary claims

There are currently six authorised claims that are subject to the protection of proprietary data, such that exclusive use of the claim has been granted to the applicant for a period of 5 years. The latest claim added to this list was for chicory inulin and 'contributes to normal bowel function by increasing stool frequency' which is discussed in this volume (Theis, 2017). Also discussed is the earlier authorised claim for reformulated, non-alcoholic, acidic (Toothkind) drinks (Sadler, 2017). A further development is that the 5-year protection period has passed for the claim for water soluble tomato concentrate, and this is in the process of being registered as a claim for general use.

1.2.1.3 Applications still pending

Following a positive EFSA opinion for an application for hydroxyanthracene derivatives relating to improvement of bowel function, concerns were raised by Member States about safety and hence this claim has not yet been authorised. Further to this, the Commission has initiated the procedure under Article 8 of the Addition of Vitamins and Minerals and of Certain Other Substances to Foods Regulation (EU, 2006b) and EFSA has been requested to provide a scientific opinion on the safety in use of this substance from all food sources.

Article 8 is the process by which a substance can be placed in one of three categories of Annex III of the Regulation (EU, 2006b); Part A: prohibited, Part B: with specified conditions, or Part C: under scrutiny for 4 years. Further to a substance being listed under Part C there is an 18-month period for submitting information in favour of the substance being permitted, and EFSA has 9 months to give its opinion on safety. Currently ephedra herb and its preparations originating from ephedra species have been added to Part A (i.e. prohibited), and yohimbe bark and its preparations originating from yohimbe have been added to Part C.

The proposed claim for hydroxyanthracene derivatives cannot be used during the period of scrutiny, effectively placing it 'on hold' for 4 years. At the end of this period the claim will either be authorised or rejected. Stakeholders were requested to submit data to EFSA by 10th June 2017 relevant to the evaluation, including technical, biological and toxicological data.

A similar development has taken place for monacolin K from red yeast rice, which has an authorised health claim for a 10 mg dose under Article 13.1 of the Nutrition and Health Claims Regulation (EU, 2006a). Though a further application was submitted under Article 13.5 for a lower dose of a proprietary product, EFSA concluded that the claim could only be used with a 10 mg dose (Le Bloc'h et al., 2015). However, the Article 13.5 application is not yet concluded. As safety concerns have also been raised about monacolin K, warning statements were discussed in relation to the Article 13.5 application, which would also apply retrospectively to the authorised Article 13.1 claim. However, a procedural issue would not allow replacing the authorised 13.1 claim with a new entry revised to include the warning statements. Hence the Commission has, on its own initiative, asked EFSA for a scientific opinion on the safety of monacolins in red yeast rice under the Article 8 procedure (EU, 2006b). EFSA's decision is expected early in 2018. The Commission has previously consulted Member States on a 10 mg dose of monacolin K. For example, Belgian guidelines recommend that red yeast rice preparations containing more than 10 mg of monacolin K/day are considered as medicinal, and preparations containing 10 mg/day and below are recommended to be considered as food supplements. The dose of monacolin K is standardised to a maximum 10% of deviation in order to avoid any undesired effects associated with higher doses.

1.2.1.4 Non-authorised applications

Five Article 13.5 claims for 'glucose' that were considered to refer to 'energy-yielding metabolism, and two claims for glyacemic carbohydrates that were related to brain or cognitive function, that were given a positive opinion by EFSA but were refused authorisation, are discussed in Chapter 15 (Arens, 2017).

1.2.2 Reduction of disease risk claims

There has been no further authorised reduction of disease risk claims since publication of Volume 2, and Table 2 in Chapter 1 of Volume 2 (pp. 5–9) provides a list of the authorised claims to date (Sadler, 2015).

1.2.3 Children's development and health claims

A further children's claim was authorised in 2016 for vitamin D and 'contribution to normal function of the immune system' for children aged 3–18 years of age, a claim that has already been authorised for the general population as a general function claim.

As of July 2017, 12 Article 14.1(b) children's development and health claims have been authorised in total (Table 1.1). A number of additional positive EFSA opinions have been issued for applications submitted by an EU trade association (SNE or Specialised Nutrition Europe, formerly called IDACE). These claims relate to micronutrients and through an individual Commission decision can still be used while awaiting authorisation. The claims are mostly for infants and children up to 3 years of age, and for most EFSA has already issued a positive opinion relating to the general population which indicates that the claims should also apply to infants and children. However, specific authorisation is required where such claims are made on products that are for the exclusive consumption of infants and children or where the claim refers solely to the development and health of children. For the Article 14.1(b) applications, evidence in children has been provided in most cases and the positive opinion has been based on this evidence plus the well-established role of the micronutrient in producing the claimed effect. However, these claims have not yet been authorised, mainly because part of EFSA's proposed conditions of use relate to compliance with compositional criteria for regulated food categories that now fall under the Foods for Specific Groups Regulation (EU, 2013). The Commission has previously indicated that authorisation of health claims for nutrients that are mandatory in regulated products for infants and young children i.e. infant and follow-on formulae (EU, 2016) and processed cereal-based food and baby food, will be considered in parallel with implementation of the Foods for Specific Groups Regulation (EU, 2013), in particular with respect to updating compositional requirements.

Where the conditions of use, relate to normal foodstuffs intended for infants and young children, EFSA has proposed that these products should provide at least 15% of the reference values (for vitamins and minerals) for nutrition labelling of such foods. However, currently the reference values (for vitamins and minerals) for nutrition labelling (nutrient reference values NRVs, or reference intakes RIs) are based on adult requirements. There has been discussion about the need to set specific NRVs and significant amounts for young children in relation to products such as young child formulae, which are now regulated as normal foods. The discussion has focused on the appropriate age range, and whether NRVs should be set for all vitamins and minerals, or for key nutrients such as iron, iodine and vitamin D. Hence, NRVs for young children are also potentially relevant for children's health claims.

The Commission is also keen that particular consideration is given to the authorised wording of these health claims, to ensure that the labelling does not mislead the purchaser as to the nutritional value of the products in comparison with other products.

Against a background that from 2020/21 all claims will be prohibited on infant formula products, the Commission has sought views on whether health claims should be allowed on regulated and non-regulated products (supplements and young child formulae) for infants and young children and if so, how the conditions of use might be determined. Indications are that the consultation responses suggest strong support to prohibit claims targeting children aged 0–3 years, and to prohibit claims on mandatory nutrients (in regulated products). However, this is not feasible within the current rules, and consideration will be needed as to how this could be achieved. Possibilities are a new recital in the legislation that would prohibit claims on mandatory nutrients in follow-on formulae, and an amendment to the voluntary claim for docosahexaenoic acid. If claims on mandatory nutrients in follow-on formulae are to be refused authorisation, it is not clear if this will also apply to un-regulated products such as dietary supplements. Progress on these claims can be followed by checking meeting summaries of the EU Standing Committee on Plants, Animals, Food and Feed (General Food Law).

1.2.4 On-hold Article 13.1 claims

There have been a number of developments in relation to five Article 13.1 claims for caffeine that were placed on hold, awaiting a safety review by EFSA because of safety concerns in children that could result in transient behaviour changes, and because moderation of caffeine intake is advisable in pregnancy. The safety review, published in May 2015, concluded that there are no safety concerns for the specific groups of the population that the panel was requested to address i.e. children and adolescents, pregnant and lactating women and adults. For children and adolescents, safe intakes are based on body weight (3 mg/kg/day). In adults, single intakes of caffeine up to 200 mg/day (about 3 mg/kg body weight for a 70 kg adult) are considered safe, even if consumed within 2 h before intense exercise. Caffeine intakes from all sources up to 400 mg/day were considered not to raise safety concerns, other than in pregnant women in whom intakes of 200 mg/day from all sources are considered safe for the developing foetus. For lactating women, single and habitual intakes of 200 mg/day from all sources are considered safe. EFSA concluded that constituents of energy drinks such as D-glucurono-gamma-lactone, taurine and alcohol would be unlikely to interact adversely with single doses of caffeine up to 200 mg. Analysis of intake data from various EU countries shows that the 95th percentile of caffeine intake from all sources exceeds these intakes in some countries for adolescents, children and toddlers.

However, these safe levels determined by EFSA were deemed to be compatible with the intakes specified in the conditions of use for four of the caffeine claims relating to alertness, concentration and for increase in endurance performance and increase in endurance capacity for caffeine taken 1 h prior to exercise. The conditions of use for a fifth claim relating to a reduction in the perceived exertion or effort during exercise, which requires 4 mg caffeine/kg taken 1 h prior to exercise, to achieve the claimed effect, are in excess of the safe level, hence this claim will be refused authorisation on these grounds.

Further to publication of the final EFSA safety review, discussions that would authorise or reject the caffeine claims were resumed. Following a detailed discussion of the target groups, safety warnings and restrictions of use, a draft Regulation was eventually agreed proposing authorisation of the following four claims:

1. caffeine helps to improve concentration: 75 mg/serving, targeting adults;
2. caffeine helps to increase alertness: 75 mg/serving, targeting adults;
3. caffeine contributes to an increase in endurance capacity: 3 mg/kg body weight taken 1 h prior to exercise, targeting adults performing endurance exercise; and
4. caffeine contributes to an increase in endurance performance: 3 mg/kg body weight taken 1 h prior to exercise, targeting adults performing endurance exercise.

The first two claims were subject to a restriction that they could not be used for foods targeting children, including adolescents, and the last two claims to a restriction that they could only be used for foods exclusively targeting adults performing endurance sports.

Further to agreement of the draft Regulation by the Standing Committee, the European Parliament (EP) in Strasbourg considered the claims under the scrutiny procedure (Binns, 2014). However, because of concerns about use of the claims on energy drinks, the EP voted to reject the draft Regulation. As well as concerns about the caffeine content of energy drinks, the EP was concerned about the high-sugar content and did not want such drinks to carry any kind of health claims—akin to imposing a nutrient profile for the sugar content of these products.

Further discussion by the Standing Committee included the proposed warnings for the claims, which it was thought may not have addressed such concerns sufficiently, and that a voluntary code of practice for marketing and labelling energy drinks has been agreed by the industry. The immediate outcome of the EP veto was that the claims are left in regulatory limbo. Nevertheless, they can still be used, as they have not officially been refused authorisation. The Commission has asked Member States to submit information about caffeine claims on energy drinks in their markets and further discussion of the claims is awaited. Regarding the fifth claim that exceeds the safety threshold, the Commission has indicated that while it is in the process of being legally refused it should not be used on foods in the EU market and should not benefit from the on-hold transitional measures.

The remaining on-hold claims include a vast number of botanical claims (discussed further), a claim for reduced lactose that will be regulated under the Food Information for Consumers Regulation (EU, 2011), and a claim for very low calorie diets that is expected to be refused authorisation (Kuczora, 2017). However, these claims have not yet been officially rejected.

1.3 REFIT evaluation of Nutrition and Health Claims Regulation

In its Better Regulation Communication in May 2015, the EU Commission announced plans to carry out a REFIT (regulatory fitness and performance programme) evaluation of the Nutrition and Health Claims Regulation. REFIT is a rolling programme that reviews EU legislation to ensure that it is 'fit for purpose'. Regarding nutrition and health claims the issues relate to certain aspects of the Regulation that remain incomplete since its adoption in 2006, mainly the setting of nutrient profiles and the incomplete regulation of claims on plants and their preparations (botanicals) used in foods. It was recognised that to address the issue of botanicals would require a broader reflection of related legislative issues.

To take this forward, in October 2015 the Commission published a roadmap outlining the purpose, content and scope of the evaluation and setting out the main criteria. These include effectiveness, efficiency, coherence, relevance and EU added value. The aims are to examine whether the current rules are adequate and how the use of 'botanical claims' interacts with the food regulatory framework on plants and their preparations. If necessary the evaluation will extend to other regulatory aspects such as safety requirements for the use of plants and their preparations in food. Terms of reference were published for an external independent study to be undertaken by a contractor, and work on this commenced in May 2016. This involves the collection of evidence and information via online surveys of Member State competent authorities, EU stakeholders, small and medium enterprises (SMEs), and an open public consultation, along with in-depth interviews and cases studies. Additionally, two workshops with Member States and stakeholders were held in 2016 that will feed into the consultation. However, the issue is unlikely to be resolved within the initially envisaged timescale as the information required to inform next steps is unlikely to be available until at least the end of 2019.

1.3.1 Nutrient profiles

The aims are to assess whether these are warranted and whether the provisions are adequate to ensure the objectives of the Nutrition and Health Claims Regulation. In its vote on the REFIT report, the EP voted in favour of a clause calling for the elimination of nutrient profiles. The vote followed a debate on 'product reformulation' initiatives by the World Health Organization to establish nutrient profiles and calls by consumer and health non-governmental organisations to take steps against health claims for foods high in fat, salt and sugar. However, consumer organisations, certain health bodies and certain multinational companies remain in favour of setting nutrient profiles.

1.3.2 Botanical claims

The aim of the roadmap is to examine whether the rules for botanical claims are adequate and how the use of botanical claims interacts with the food-regulatory framework on plants and their preparations. If necessary the evaluation will extend to other regulatory aspects, such as safety requirements. The issue is complex with a mismatch between the requirement for clinical studies in the Nutrition and Health Claims Regulation, which addresses foods, and the acceptance of 'traditional use' evidence for therapeutic indications in medicinal products. Solving this inconsistency is a major challenge and is seen as a matter of priority.

1.4 Communications to health professionals

In 2016, the Advocate General gave an opinion on a German court case requesting clarification about the application of the Nutrition and Health Claims Regulation (EU, 2006a) to business communications targeted at health professionals. The Advocate General concluded that the Regulation should apply to communications of this sort, as their goal is to reach consumers indirectly via these channels. The Commission and the French and Greek governments have supported this position. The Advocate General concluded that the key issue is whether the foods themselves are intended for the final consumer, and not the groups to which the communications are targeted. The Advocate General's opinion is not binding on the Court but is highly influential should there be a need for any future judgements on such issues.

1.5 Guidance

1.5.1 Guidance for SMES

A best practice guide for intervention studies has been published (Lucey et al., 2016) specifically for the purpose of EU health claim applications. The guide is part of a toolkit that has been developed through the EU-funded BACCHUS project, to support the needs of SMEs that are considering applying for EU health claims. Other parts of the toolkit include an e-learning platform, a bioactive database, an intake assessment tool and more best practice guides. Though the toolkit was developed for use by SMEs, it will undoubtedly be of value to other stakeholders such as research scientists, students and advisers.

1.5.2 Revisions to EFSA Guidance

In January 2017, EFSA published a second revision of its scientific and technical guidance for the preparation and presentation of health claim applications (EFSA, 2017), updating the previous version published in 2011. This followed a consultation in July 2016 to which EFSA received comments from certain companies operating in the food industry, food industry associations, consultants, a research consortium and a non-governmental organisation. Many of these comments were outside the objectives and scope of the draft guidance including a number of comments about the legal framework for the authorisation of health claims, and those relating to risk-management aspects, which were beyond EFSA's remit and could not therefore be taken into account.

As with all such guidance the revision is based on lessons EFSA has learned from assessing applications to date. Though much of the information required for the supporting dossier remains as before, it nevertheless sets out a more logical format for organising the required information than was previously the case. Clearer guidance is given on the distinction between substantiating and supporting data and there are fewer demands in terms of tabulating the identification of different study types during the systematic review process. A key revision is the addition of advice on the distinction between claims based on the essentiality of nutrients and those for other substances, for which the data requirements are different for characterisation of the nutrient, for characterisation of the claimed effect, for scientific substantiation of the claim, and for establishing the conditions of use. Advice for characterisation of the substance now includes vitamins and minerals and fixed combinations of vitamins and minerals. There is also a requirement to include a section in the application dossier about characterisation of the claimed effect. For reduction of disease risk claims [Article 14.1(a)] there is advice about characterising both the risk factor and the disease, which was previously lacking. The application must now include details of the process that was used to find unpublished studies during the systematic review process, and advice is given for reporting unpublished or proprietary studies. The latter includes a template for writing up the study, or alternatively guideline E3 of the International Conference on Harmonisation (ICH) on the structure and content of clinical study reports can be followed (ICH, 1995).

Prior to this in January 2016, EFSA published revised general guidance on health claims (EFSA, 2016a), updating the previous version published in March 2011. This sets out the general scientific principles applied when EFSA assesses health claims. The revision is more comprehensive than the previous version, providing more in-depth information about certain aspects of the scientific assessment and giving a broad-brush indication of what is required in an application. Particular sections outline the type of evidence required to determine whether a chosen risk factor for a reduction of disease risk claim is an 'independent risk marker', and how data in meta-analyses is assessed, which has previously been a point of controversy.

EFSA is also in the process of revising its guidance relating to specific areas of health claims and has so far addressed the guidance relating to antioxidants, oxidative damage and cardiovascular health, and to the immune system, gastrointestinal effects and defence against pathogens (EFSA, 2016b). The latter addresses more examples of claims relating to gastrointestinal effects than previously included e.g. sections on gastrointestinal discomfort claims, for both infants and for the general population; and addressing claims on the digestion and absorption of both macronutrients and micronutrients. The guidance also addresses more claims relating to the immune system, with specific advice for claims targeting the respiratory, lower urinary and gastrointestinal tracts. The revision also provides guidance on disease risk–reduction claims in the area and addresses how claims for essential nutrients and other substances are assessed differently.

From these revisions it is clear that EFSA guidance is becoming more comprehensive as more experience is gained in evaluating health claims. However, the underlying scientific principles behind health claim evaluations remain unchanged, and the data requirements are no less demanding. Also the documents do not address all potential health relationships such that it will remain difficult to predict the outcome of future health claim submissions if the proposed claims are based on health relationships not previously considered by EFSA.

1.6 EFSA developments

1.6.1 Changing responsibilities of panels

A Commission Regulation published in February 2017 (EU, 2017) will change the name and the work areas of 3 of the 10 EFSA scientific panels. This reflects expected changes in scientific and technical developments that will impact the workload of these panels, including an expected decrease in health claim applications and work on dietary reference values, an expected increase in the need for evaluation of nutrient sources and other substances with physiological effects added to food, and an expected increase in evaluations of pending applications for food enzymes. The NDA Panel (Dietetic products, Nutrition and Allergies), that assesses health claims, will be changed to 'Panel on Nutrition, Novel Foods and Food Allergens', the ANS Panel (Food Additives and Nutrient Sources) will be changed to 'Panel on Food Additives and Flavourings', and the CEF Panel (Food Contact Materials, Enzymes, Flavourings and Processing Aids) will be changed to 'Panel on Food Contact Materials, Enzymes and Processing Aids'. Evaluation of nutrient sources and other substances will be assigned to the NDA panel, in view of the reduced workload in relation to dietary reference values and health claims. The evaluation of flavourings, currently undertaken by the CEF panel will be assigned to the ANS panel. These changes will not take place until 1st July 2018, allowing time for reorganisation. This suggests that the currently reduced number of health claim applications being submitted is expected to continue in the longer term, and that the Nutrition and Health Claims Regulation is having a major impact on industry innovation, which is discussed in Chapter 2 (Hieke and Grunert, 2017).

1.7 Changing political landscape

1.7.1 Potential impact on Brexit of regulatory issues relating to health claims

It is widely recognised that the process of the United Kingdom leaving the European Union (Brexit) will not be easy and will take some considerable time. In terms of food products and food ingredients, there is significant trade between EU Member States and the United Kingdom, which in principle benefits from harmonised laws and legislation. However, the full implications, particularly in relation to food law and the regulation of health claims, will not become clear until a decision is made on what type of relationship will exist between the United Kingdom and the European Union post Brexit, as this will determine the scope by which future UK food regulation may become divergent from that in the European Union e.g. if a position may be negotiated that would not require the United Kingdom to maintain the current EU Nutrition and Health Claims Regulation.

For businesses there are both potential, advantages and disadvantages of any future divergence for health claims. On the down side this will create new costs and complexities for businesses that sell products bearing health claims in both the European Union and the United Kingdom, as they will need to straddle two sets of regulations. However, a different approach to health claims might create some opportunities for the UK domestic market, particularly in terms of innovation. If new UK rules make health claims less burdensome than at present, this could have a positive impact for companies wishing to sell products with health claims in the United Kingdom, which could particularly benefit SMEs.

Initially the UK government plans to transpose EU law into UK law, to ensure minimal changes for industry, which will also mean minimal changes for consumers. Any opportunity for the United Kingdom to amend its own laws and to develop new ones will only become apparent after the country has exited the European Union. One such area may indeed be the regulation of health claims, not least because the United Kingdom will no longer be subject to the advice and opinions of EFSA, which is a European, independent advisory body. The United Kingdom will thus need its own source of independent scientific advice, whatever form this may take. However, as EFSA is so pivotal to the Nutrition and Health Claims Regulation, having to put alternative arrangements in place could create an opportunity for change in the United Kingdom. A further potential reason for divergence relates to the REFIT process, particularly for the issues of nutrient profiles and the future regulation of botanicals. If Europe agrees changes to the Nutrition and Health Claims Regulation before the United Kingdom exits the European Union, then these will likely be implemented in the United Kingdom, but if Europe agrees changes after the United Kingdom has exited the EU, then this could result in immediate divergence. Longer term is the issue of trademarks and brand names existing before 2005 that have a transition period until 19 January 2022 to comply with the Regulation. As this is after the projected date for the United Kingdom to leave the European Union, this issue will require some form of resolution.

Before the Nutrition and Health Claims Regulation was adopted, the United Kingdom was foresighted and adopted its own voluntary system of health claims oversight. The Joint Health Claims Initiative (JHCI) was a tripartite agreement between enforcement, industry and consumers. Its remit was to offer pre-market advice, and a code of practice was developed to ensure that health claims on foods and in marketing were both scientifically true and legally acceptable. The code was administered jointly by the JHCI Council, the Secretariat, and an Expert Committee of independent scientific experts who assessed the scientific evidence submitted by companies wishing to make a health claim. The code was applicable to any health claim made in advertising, marketing including the Internet, and product labelling, and worked alongside the food legislation existing at that time. JHCI produced guidelines to assist industry in preparing dossiers to substantiate health claims. Consulting the JHCI and following the code was considered an important step in developing all due diligence should a dispute arise. It also helped to ensure consumer confidence in the truthfulness of the claim as it was assessed by expert scientists. JHCI also worked to develop well-established nutrient function statements for the industry to use, which were developed at the request of the UK Food Standards Agency.

However, as the JHCI service was voluntary, it was not much used by the industry. A further problem was that, once claims were approved, there was no distinction between the approved claims and the claims that had not been through the process; particularly as in some cases applications were kept confidential. Interestingly, JHCI approved claims that were subsequently refused authorisation in the European Union e.g. for whole grains and heart health, and for soya protein and reduction of blood cholesterol, having been subject to significant scientific scrutiny. JHCI allowed claims to be based on evidence from observational studies, provided the wording reflected this, rather than having to demonstrate 'cause and effect' as in the more 'pharmaceutical'-type model adopted by EFSA. This suggests that there is scope for alternative approaches that are acceptable between consumers, enforcement and industry, and that are nevertheless rigorously scientific.

To underline this point, other systems of health claims regulation exist in non-EU countries, and have more flexible approaches than that in the European Union. For example, Food Standards Australia and New Zealand (FSANZ) approved a new food standard (1.2.7) on nutrition, health and related claims in January 2013, and companies had until January 2016 to comply (FSANZ, 2013). This requires, that health claims must be based on food–health relationships that have been substantiated according to the FSANZ standard. 'High-level' health claims must be based on food–health relationships pre-approved by FSANZ. However, for 'general level' health claims the standard includes more than 200 pre-approved food–health relationships that companies can use. Additionally, there is scope for self-substantiation of 'general level' food–health relationships by following the process for systematic review described in the code. Self-substantiated food–health relationships must be notified to FSANZ before the claim is made on food labels or in advertising. Whilst there is no further scrutiny of the claim, a list of notified relationships is maintained as a public record. Notified claims are for the sole use of the company submitting the claim, such that if a different food business wants to make the same claim they must conduct their own review of the evidence. This system thus incorporates elements of self-regulation allowing for a more flexible approach than the system adopted in Europe, but nevertheless places the burden on industry to be rigorous in any claims they are making. FSANZ does not provide advice to businesses using the code, such that advice may need to be sought from consultants or lawyers, or both.

Another example is Switzerland, which has its own system for approving health claims and a number of health claims for prominent, branded probiotic products have been approved by the Federal Food Safety and Veterinary Office (FSVO), suggesting a more flexible system than that adopted in the European Union, where no claims for probiotics have yet been approved.

A less burdensome regulatory regime for health claims in the United Kingdom would thus not be out of line with other countries, and could be advantageous to both, industry and consumers.

Other issues, relating to potential divergence in health claims regulations, are that previous United Kingdom influence on EU laws will no longer exist. Hence developing regulatory issues in Europe that are relevant to health claims will be beyond the United Kingdom's control. More generally, for situations where the United Kingdom was previously at odds with other Member States on particular issues, the European Union can move forward on such issues in the future in the absence of UK influence. Hence United Kingdom based companies that sell into the European Union may want to maintain their own influence over the future development of health claims regulation in Europe, and will need to plan for this in the longer term by developing their contacts and networks.

A further consideration is the issue of health claim applications. Clearly in the future should UK companies wish to sell products with new health claims in the European Union, they will still need to apply for EU authorisation, and applications will need to be submitted via the competent authority in an EU Member State rather than via the United Kingdom. While the United Kingdom currently remains in the European Union, health claims can still be submitted via the UK. However, businesses applying for health claims in the short-term may decide to use an alternative Member State. It is well documented that the majority of health claim applications take longer than 2 years from submission to the Member State through to authorisation or refusal, and if the application is not completed by the time the United Kingdom leaves the European Union, this could add complexity to the process.

1.7.1.1 Potential impact on UK food research

EU funds for the UK research activity are considerable, including in the area of food research, by way of numerous collaborative projects. Hence the funding of studies conducted in the United Kingdom and investigating nutritional and physiological effects that are potentially relevant to substantiating health claims is an area that could be negatively impacted by Brexit.

1.8 Conclusions

EU Regulation 1924/2006 on Nutrition and Health Claims Made on Foods was published in 2006, and though the Regulation is by now well bedded-in, food business operators continue to operate in a climate of uncertainty on various issues, including the use of numerous proposed children's claims for micronutrients, nutrient profiles and health claims for botanicals. The latter two issues are subject to a wide-ranging review within the Commission. EFSA has published revised guidance to assist with health claim applications. However, the remit of the expert panels is being revised, reflecting the reduced number of health claim applications. Consideration of the potential impact of Brexit on health claims suggests that this may provide a future opportunity for divergence in regulation depending on the final relationship that is negotiated between the United Kingdom and the European Union.

1.9 Sources of further information and advice

Addition of Vitamins and Minerals Regulation Information (https://ec.europa.eu/food/safety/labelling_nutrition/vitamins_minerals_en).

EFSA Fact Sheet on Caffeine Risk Assessment (http://www.efsa.europa.eu/sites/default/files/corporate_publications/files/efsaexplainscaffeine150527.pdf).

EFSA Register of Questions (http://registerofquestions.efsa.europa.eu/roqFrontend/ questionsListLoader?unit=NUTRI).

EU Regulations (http://ec.europa.eu/food/food/labellingnutrition/index_en.htm).

EU Register on Nutrition and Health Claims (http://ec.europa.eu/food/safety/labelling_nutrition/claims/register/public/?event=register.home).

Food Standards Australia New Zealand (FSANZ) Health Claims (http://www.foodstandards.gov.au/industry/labelling/Pages/Nutrition-health-and-related-claims.aspx).

Standing Committee on Plants, Animals, Food and Feed (General Food Law). (https://ec.europa.eu/food/safety/reg_com/gfl_en).

Switzerland Federal Food Safety and Veterinary Office (FSVO) (https://www.blv.admin.ch/blv/en/home.html?_organization=1079).

References

Arens, U., 2017. Authorised EU health claim for carbohydrates and maintenance of normal brain function. Sadler, M.J. (Ed.), Foods, Nutrients and Food Ingredients with Authorised EU Health Claims, vol. 3, Woodhead Publishing, Cambridge, (Chapter 15).

Binns, N., 2014. The regulation of health claims in Europe. Sadler, M.J. (Ed.), Foods, Nutrients and Food Ingredients with Authorised EU Health Claims, vol. 1, Woodhead Publishing, 3–22, (Chapter 1).

Chappuis, E., 2017. Authorised EU health claim for glycaemic carbohydrates and muscular recovery. Sadler, M.J. (Ed.), Foods, Nutrients and Food Ingredients with Authorised EU Health Claims, vol. 3, Woodhead Publishing, Cambridge, (Chapter 9).

EFSA, 2016a. In: General scientific guidance for stakeholders on health claim applications. EFSA J. 14 (1), 4367.

EFSA, 2016b. In: Guidance on the scientific requirements for health claims related to the immune system, the gastrointestinal tract and defence against pathogenic microorganisms. EFSA J. 14 (1), 4369.

EFSA, 2017. Scientific and technical guidance for the preparation and presentation of a health claim application (Revision 2). EFSA J. 15 (1), 4680.

EU, 2006a. Regulation (EC) No. 1924/2006 of the European Parliament and of the Council of 20 December 2006 on nutrition and health claims made on foods. Off. J. Eur. Union 30.12.2006, L404, 9–25.

EU, 2006b. Regulation (EC) No. 1925/2006 on the addition of vitamins and minerals and of certain other substances to foods. Off. J. Eur.Union 20.12.2006, L404, 26–38.

EU, 2011. Regulation (EU) No. 1169/2011 of the European Parliament and of the Council of 25 October 2011 on the provision of food information to consumers, amending Regulations (EC) No. 1924/2006 and (EC) No. 1925/2006 of the European Parliament and of the Council, and repealing Commission Directive 87/250/EEC, Council Directive 90/496/EEC, Commission Directive 1999/10/EC, Directive 2000/13/EC of the European Parliament and of the Council, Commission Directives 2002/67/EC and 2008/5/EC and Commission Regulation (EC) No. 608/2004. Off. J. Eur. Union 22.11.2011, L304, 18–63.

EU, 2013. Regulation (EU) No. 609/2013 of the European Parliament and of the Council of 12 June 2013 on food intended for infants and young children, food for special medical purposes, and total diet replacement for weight control and repealing Council Directive 92/52/EEC, Commission Directives 96/8/EC, 1999/21/EC, 2006/125/EC and 2006/141/EC, Directive 2009/39/EC of the European Parliament and of the Council and Commission Regulations (EC) No. 41/2009 and (EC) No. 953/2009. Off. J. Eur. Union 26.6.2013, L181, 35–86.

EU, 2016. Commission Delegated Regulation (EU) 2016/127 of 25 September 2015 supplementing Regulation (EU) No. 609/2013 of the European Parliament and of the Council as regards the specific compositional and information requirements for infant formula and follow-on formula and as regards requirements on information relating to infant and young child feeding. Off. J. Eur. Union 2.2.2016, L25, 1–29.

EU, 2017. Commission Regulation (EU) 2017/228 of 9 February 2017 amending Regulation (EC) No. 178/2002 of the European Parliament and of the Council as regards the names and the areas of competence of the scientific panels of the European Food Safety Authority. Off. J. Eur. Union 10.2.2017, L035, 10–11.

FSANZ, 2013. Australia New Zealand Food Standards code—Standard 1.2.7—Nutrition, health and related claims. (https://www.legislation.gov.au/Details/F2017C00310).

Harland, J., 2017. Authorised EU health claim for sugar beet fibre. Sadler, M.J. (Ed.), Foods, Nutrients and Food Ingredients with Authorised EU Health Claims, vol. 3, Woodhead Publishing, Cambridge, (Chapter 8).

Hieke, A., Grunert, K.G., 2017. Consumers and health claims. Sadler, M.J. (Ed.), Foods, Nutrients and Food Ingredients with Authorised EU Health Claims, vol. 3, Woodhead Publishing, Cambridge, (Chapter 2).

ICH, 1995. Structure and content of clinical study reports E3. ICH Expert Working Group Report, 30 November 1995.

Kuczora, S., 2017. Authorised EU health claim for meal replacements. Sadler, M.J. (Ed.), Foods, Nutrients and Food Ingredients with Authorised EU Health Claims, vol. 3, Woodhead Publishing, Cambridge, (Chapter 12).

Le Bloc'h, J., Pauquai, T., Bourges, C., 2015. Sadler, M.J. (Ed.), Foods, Nutrients and Food Ingredients with Authorised EU Health Claims, vol. 2, Woodhead Publishing, 139–152, (Chapter 8).

Lucey, A., Heneghan, C., Kiely, M.E., 2016. Guidance for the design and implementation of human dietary intervention studies for health claim submissions. Nutr. Bull. 41 (4), 378–394.

Sadler, M., 2015. Regulatory developments with European Health Claims. Sadler, M.J. (Ed.), Foods, Nutrients and Food Ingredients with Authorised EU Health Claims, vol. 2, Woodhead Publishing, Cambridge, 1–28, (Chapter 1).

Sadler, M., 2017. Authorised EU health claim for reformulated, non-alcoholic, acidic drinks. Sadler, M.J. (Ed.), Foods, Nutrients and Food Ingredients with Authorised EU Health Claims, vol. 3, Woodhead Publishing, Cambridge, (Chapter 7).

Shortt, C., 2014. Authorised EU health claims for intense sweeteners and sugar replacers. Sadler, M.J. (Ed.), Foods, Nutrients and Food Ingredients with Authorised EU Health Claims, vol. 1, Woodhead Publishing, Cambridge, 151–176, (Chapter 8).

Theis, S., 2017. Authorised EU health claim for chicory inulin. Sadler, M.J. (Ed.), Foods, Nutrients and Food Ingredients with Authorised EU Health Claims, vol. 3, Woodhead Publishing, Cambridge, (Chapter 10).

Further Reading

EU, 2012. Commission Regulation (EU) No. 432/2012 of 16 May 2012 establishing a list of permitted health claims made on foods, other than those referring to the reduction of disease risk and to children's development and health. Off. J. Eur. Union 25.5.2012, L131.

Harland, J., 2014. Authorised children's development and health claims in the EU. Sadler, M.J. (Ed.), Foods, Nutrients and Food Ingredients with Authorised EU Health Claims, vol. 1, Woodhead Publishing, 73–91, (Chapter 4).

Consumers and health claims

2

Sophie Hieke, Klaus G. Grunert***
*European Food Information Council, Brussels, Belgium; **MAPP Centre, Aarhus University, Aarhus C, Denmark

2.1 Introduction

Originally, health claims were designed as marketing tools to highlight health benefits of a product to consumers and impact food choice of a growing segment of health-conscious shoppers. After their regulation in 2006 (EC 2006), the use of health claims has been harmonised by means of a scientific substantiation and validation process. The Regulation was designed to ensure that claims protect consumers and their right to non-misleading food information, that they provide useful and reliable information and that they are to be understood by the average consumer. It was also meant as a guideline for industries on how to use claims on food products, ensuring the effective functioning of the internal EU market and as such fostering innovation potential and competitiveness of the European food sector (EC, 2006).

More than a decade later, the transition period has been completed, all foods *should* comply with the legislation, and consumers should be familiar with such health-related information on their food products. The reality is that the actual role that health claims play in consumer behaviour is not clear. Food choice, purchase and consumption behaviour are complex cognitive and often emotionally driven processes, layered in multiple contexts such as culture, country history, socio-demographics, individual preferences and even situational factors (appetite, time pressure, mood or simply a sick child at home). A multitude of research studies has been conducted to better understand these factors, including a pan-European research project on the role of claims and symbols in consumer behaviour (CLYMBOL, No. 311369, www.clymbol.eu). A brief overview of the main findings will lay the ground for the work presented in this book and the two previous volumes, and will paint the first picture of the current knowledge we have on the effectiveness of health claims on the European market.

CLYMBOL (deriving its name from 'the role of health-related CLaims and sYMBOLs in consumer behaviour') was an EU-funded project aiming to study how health claims and symbols influence consumer understanding, purchase and consumption behaviour. Health claims and symbols were studied in their context, for example as they appear on-pack, together or without additional (visual) information and how they interplay with national (cultural) differences as well as personal factors such as motivation and/or ability to process this health-related information. During a 4-year period, a wide range of research studies have been conducted across Europe. Results were intended to provide a basis for recommendations for several stakeholders such as policy makers, the food industry and consumer and patient organisations (Hieke et al., 2015).

Foods, Nutrients and Food Ingredients with Authorised EU Health Claims. http://dx.doi.org/10.1016/B978-0-08-100922-2.00002-4
Copyright © 2018 Elsevier Ltd. All rights reserved.

2.2 Regulation of health claims in Europe and food innovation

In 2000, the European Commission first regulated the labelling, presentation and advertising of foodstuffs by prohibiting claims for prevention, treatment or cure of human diseases on food products (Directive 2000/13/EC). However, this legislation was not consistently enforced throughout the European Union (Bech-Larsen and Scholderer, 2007). Further initiatives, among others the concerted action projects such as Functional Food Science in Europe (FUFOSE) and Process for the Assessment of Scientific Support for Claims on Foods (PASSCLAIM) that developed draft guidelines for assessing the scientific substantiation of health claims (Asp and Bryngelsson, 2008; Richardson, 2012), were needed to pave the way for the supranational Regulation (EC) No. 1924/2006 on the use of nutrition and health claims, thus harmonising the law in Europe.

To use a health claim on a food product, food businesses can either check existing lists for available health claims and their conditions of use or apply for the authorisation of a new health claim. The Nutrition, Dietetic Products and Allergies (NDA) panel of the European Food Safety Authority (EFSA) carries out this scientific assessment of evidence. All health claims must be substantiated by generally accepted scientific evidence. The European Commission bases its final decision on whether to approve or reject each claim on EFSA's scientific opinion, together with considerations of consumer understanding of the claim (Binns, 2014; Buttriss, 2015).

The Regulation (EC) No. 1924/2006 has resulted in an established list of health claims that can be used on food and drink products, given that they meet the conditions of use set during the authorisation procedure. To date, over 3000 health claims have been submitted to EFSA, of which 266 have been evaluated with a positive outcome so far. They can be accessed in the EU register on nutrition and health claims (http://ec.europa.eu/food/safety/labelling_nutrition/claims/register/public/?event=register.home). Some of the submitted claim dossiers were found to show insufficient evidence, while the large majority of health claims proposed was considered not to be supported by scientific evidence (Verhagen and van Loveren, 2016).

The implications of this Regulation have been widely discussed. Uncertainty over the outcome of a claim evaluation and lack of transparency in how EFSA evaluates claim dossiers have been mentioned (Buttriss and Benelam, 2010). It has been further argued that the Regulation places a burden on food manufacturers; specifically, the costs for research and development, reformulation, marketing and regulatory affairs are seen as high (Leathwood et al., 2007; Van Buul and Brouns, 2015). In 2010, they were estimated to be between 0.26 and 1 million EUR (Brookes, 2010). Such a burden is particularly high for small- and medium-sized enterprises: in a survey in 2015, carried out by the Impact of Biogenic versus Anthropogenic emissions on Clouds and Climate: towards a Holistic UnderStanding (BACCHUS) project, less than a third reported considering submitting a health claim dossier, citing financial costs of the process as the main barrier, followed by the length of time and complexity of the process as well as an uncertain likelihood of success (Buttriss, 2015).

Another aspect in this discussion has been the innovation and competitiveness of the European food sector that the Regulation intended to foster (Aschemann-Witzel, 2011).

Current literature suggests several challenges that the Regulation may pose for food innovation: a focus on ingredients that are linked to health claims could lead to diminished investments in innovative food science and technology and thus support incremental innovation, rather than actual innovation. Lack of transparency regarding the requirements for a successful health claim dossier could lead to companies reducing their research and development expenditures. Uncertainty over the flexibility in wording of claims as well as limited financial/R&D resources are further cited as potential barriers to innovation in functional foods. Lastly, in more extreme cases, it has been argued that companies that do not obtain approval for health claims for their products may find it difficult to maintain their products in the same category and may be forced to switch product categories or even abandon the functional foods market altogether (Bröring et al., 2016). The authors argue that when a discouraged product differentiation based on health claims results in 'me-too' products (using approved claims that were submitted by other companies), new product development and radical innovation could be replaced by a supply of products having similar features and simple reformulations (incremental innovation). Consequently, in a 2014 survey of companies involved in the EU food sector, it was found that wording of claims, missing transparency and limited financial resources are major challenges in complying with the Regulation and companies reported not to have increased their R&D expenditure or innovation activities since the implementation of the Regulation (Khedkar et al., 2016a). While the authors conclude that, currently, the Regulation does not seem to have fostered innovation in the EU food sector, they also point at a potential positive indirect effect on product innovation, especially in the case of small- and medium-sized enterprises: by increasing the demand for external sources of knowledge, enhancing networking capacities of smaller companies, supporting collaboration and ultimately impacting internal product and process innovation (Khedkar et al., 2016b).

2.3 Health claims: what is currently on the market?

A multitude of food, health and well-being messages can be found in the food landscape. For a meaningful and structured analysis of their role in consumer behaviour, it is pertinent to develop a classification scheme, defining and grouping various pieces of food labelling information and ensuring a common understanding of the terminology used.

Such a taxonomy has been developed by CLYMBOL which is based on EU regulations as well as Codex Guidelines. The main distinction made regarding food labelling is that of *information* versus *claims*. Information can be mandatory [e.g., by the Food Information to Consumers Regulation EU (2011)], or other (e.g. value-chain related information such as production methods (organic) or trade conditions (fair trade), marketing-related information such as brands or trademarks, package design elements and additional forms of presenting nutrition information such as a repetition of ingredients on-pack. Claims, on the other hand, represent any message, including pictorial, graphic or symbolic representation which states, suggests or implies that a food has particular characteristics. CLYMBOL focussed on expressed (vs. implied) claims which could be either worded (written claims) or symbolic (symbols and logos). Following the Regulation, a distinction was made between nutrition and health claims, and consequently nutrition and health symbols. A third category, health-related ingredient claims, was introduced to cover all statements referring to ingredients or food constituents which do not fall under 'other substances' (Article 2.2 of the Regulation), for example '100% whole grains' or 'one of your 5-a-day'. These can also be either worded or symbolic. Any other claims and symbols not fitting into any of these categories were classified as miscellaneous.

With this taxonomy at hand, all information on-pack can now be classified, using a shared terminology. Each category can be further refined [e.g. health claims can be classified as per their ingredient, general function or outcome; their effect direction (increase, reduction, maintenance etc.), certainty (with or without a qualifier such as helps, contributes etc.) or framing (positive vs. negative); or as per aspects such as target group, level of scientific agreement and type of data that the claimed effect is based on (generally accepted vs. newly developed scientific evidence)]. Using such a classification scheme, the type of information and claims found on food labels can be studied, analysed and compared across countries.

It has been argued that the Regulation would not lead to fewer claims on the market, but rather remove unfounded claims while simultaneously adding greater credibility to the remaining ones (Verhagen et al., 2010). Prevalence studies are a suitable tool to measure and monitor the development on the food market and create data on the types of products and health claims that consumers are exposed to.

CLYMBOL conducted such a multinational survey (across Germany, Spain, the Netherlands, Slovenia and the United Kingdom) in 2013, on more than 2000 food products, and found that between 20% and 35% of food products carried a claim. Nutrition claims were the most frequently found claims, followed by health claims and health-related ingredient claims (non-nutrient substances which may have a nutritional or physiological effect). UK prepackaged food products had the highest prevalence of nutrition claims, whereas products in the Netherlands had the most health claims. Most of these were the Dutch Choices logo that CLYMBOL classified as a symbolic general health claim. Almost half of all health claims found were nutrient and other function claims (47%), while only 5% were disease risk reduction claims. The highest prevalence of claims was found on baby foods (78% carried a nutrition claim and 71% carried a health claim), compared with convenience foods (9% carried a nutrition claim) and egg products which did not carry any claim. The survey also showed that foods tend to carry more than one claim: either the same claim was repeated several times on the product package or a health claim was accompanied by a related nutrition claim (Hieke et al., 2016). Considerable differences as compared with previous research could be demonstrated, with older European studies generally reporting higher percentages of nutrition and health claims on foods. Serbia, as a current candidate to join the European Union and as such not yet governed by the Nutrition and Health Claims Regulation, reported far less claims on foods.

CLYMBOL further analysed the nutritional quality of these sampled food products, comparing products with and without health claims. Results showed that foods carrying health-related claims had marginally better nutrition profiles than those that did not carry claims. In particular, foods with health claims were lower in energy, protein, total sugars, saturated fat and sodium, and higher in fibre than foods without a claim. To date, however, it is still unclear whether these relatively small differences can have a significant impact on public health (Kaur et al., 2016).

The data collected in the multi-country survey were also analysed for their packaging, to better understand in what context health claims appeared on foods. While no dominant packaging type for products carrying claims could be found across all countries, cartons were most often used for claim-labelled products in Germany, the Netherlands and Slovenia, compared with multipacks in Spain and flexible packaging in the United Kingdom.

Concerning the imagery on food packages carrying claims, the most prevalent ones found were images of the food itself or images of fresh ingredients contained in the product. In some food groups, every single product carrying a claim also displayed images of the food itself. Images of other food(s) were also very popular, for example on meat products. Images depicting nature were most popular in Spain (almost all food groups), followed by the United Kingdom. In the other countries, only specific food groups such as dairy had a strong prevalence of nature imagery for products carrying claims. Images of persons (e.g. people, children, slim figures, physically active people or cartoons) were less widespread, with large country and food group differences. Unsurprisingly, many baby foods across countries carried pictures of babies, infants and toddlers. In Germany, dairy products often showed physically active people, while in the Netherlands, images of people in general were rather popular, across various food groups. In Slovenia and Spain, although far less in frequency, cartoons were found most often on food packages carrying claims. In the United Kingdom, while images of people were more popular, no clear tendencies concerning food groups nor types of imagery could be identified.

Most claims identified in this study were found front-of-pack. This is not surprising as research has shown front-of-pack to be more effective in driving recall of health claims (Van Kleef et al., 2008). However, claims were rarely placed on the inside of a pack or on the top. The most popular form of presenting claims was in text, followed by texts and graphics. Claims were typically written in small size, competing with a number of other pieces of information contained on the product packaging.

2.4 How health claims affect consumers

A number of studies have investigated how consumers perceive health claims, their attitudes to them and whether they understand them correctly. Studies have also addressed whether health claims play a role in consumer purchase decisions, and whether they have an impact on consumption. We will in the following sections summarise major results from these lines of research.

2.4.1 Perception and attitudes

In a framework developed in 2012 for analysing consumer reactions to functional food, the determinants of the effects of health claims on consumer attitudes and perception were separated into two categories: product-related variables and consumer-related variables. *Product-related variables* include the type of carrier product, health claim format, wording of claims, functional ingredient/benefit claimed and taste/sensory attributes (Wills et al., 2012). For example product categories that have a healthy image are seen as more suitable to carry health claims, although this can differ by country. The type of claim can further impact consumer perceptions: studies show different responses to different types of claims (general health claim, nutrient and other function claim or disease risk reduction claim), depending not only on whether the ingredient is mentioned, but also on whether the claim is framed positively or negatively. To make matters more complicated, results can differ by country (Van Wezemael et al., 2014; Wills et al., 2012). Results on the effects of the wording of claims are mixed, with a newer literature review finding that it has little impact on how claims are perceived (Lähteenmäki, 2013). The biggest impact on claim perception was found for the ingredient/benefit claimed and consumers' familiarity with it (Lähteenmäki, 2013; Miklavec et al., 2015; Wills et al., 2012). Lastly, taste expectation and sensory attributes have been shown to influence the perception of products with health claims, with consumers generally not willing to comprise taste for health (Wills et al., 2012). Following the healthy = less tasty intuition, it is argued that consumers can link products perceived as healthy to a lower taste (Raghunathan et al., 2006) and, consequently, use of health claims can be lower for consumers who state taste as an important driver in their food choice (Orquin and Scholderer, 2015; Petrovici et al., 2010; Vadiveloo et al., 2013). *Consumer-related variables* include personal beliefs, personal relevance, familiarity and experience and nutrition knowledge. Attitude towards functional foods has been shown to have a significant effect on the perception of health claims and health claim-labelled products, as do personal relevance and previous experience with functional foods (Wills et al., 2012). Some studies even showed that the effect is strongest when the claim is relevant to self (Dean et al., 2012; Wong et al., 2013). Evidence of the role of nutrition knowledge is mixed, with some studies showing positive effects while other studies did not (Cavaliere et al., 2016; Lalor et al., 2011; Wills et al., 2012).

2.4.2 Understanding

The Regulation states that claims must be understood by the average consumer, defining the average consumer as reasonably well informed, observant and circumspect. Claims shall not be false, ambiguous or misleading; give rise to doubt about the safety and/or nutritional adequacy of other foods; encourage or condone excess consumption of food; state, suggest or imply that a balanced and varied diet cannot provide appropriate quantities of nutrients in general; and/or refer to changes in bodily functions which could give rise to exploit fear in the consumer (EC, 2006).

Research into consumer understanding of health claims is still limited. A few studies have investigated subjective understanding, that is if consumers themselves thought that they understood the claim, showing high (Lynam et al., 2011) or moderate (van Trijp and van der Lans, 2007; Wong et al., 2014) levels of self-reported understanding. However, believing to have understood the claim does not necessarily mean that the understanding is objectively correct. In analysing possible misinterpretations of health claims, two lines of research have been followed. The first deals with specific types of misunderstanding, most notably *the halo effect* and the *magic bullet effect*. The halo effect implies that consumers extend the benefit expressed in the claim to other product attributes, thus over-interpreting the benefits that come with the product by, for example inferring that a product with a health claim is generally superior in terms of its nutritional qualities. The magic bullet effect implies a similar strategy, with consumers attributing inappropriate general health benefits to the product, based on the presence of a health claim. Some studies have found evidence for such a halo effect (Andrews et al., 1998; Harris et al., 2011), while others have not (Lähteenmäki et al., 2010; Orquin and Scholderer, 2015), and also the evidence on the occurrence of magic bullet effects is mixed (Bilman et al., 2012; Lähteenmäki et al., 2010; Roe et al., 1999).

Only few studies have actually tried to assess whether consumer understanding is compatible with the scientific dossier on the basis of which the health claim was approved, even though this is the gist of the EU Regulation on adequate understanding of health claims. Grunert et al. (2011) used an open-ended approach developed by Danone (Rogeaux, 2010) to assess German consumers' understanding of a health claim on yoghurt, and found that most consumers in the study could be classified as 'safe' in their interpretation of the claim, that is their interpretation was in line with the scientific dossier. However, some consumers also made risky interpretations, and consumers with a positive attitude to health claims were more likely to make such risky interpretations. In a study conducted by Stancu (2016) in Denmark and Italy on health claims on olive oil, it was likewise found that most interpretations were safe, although there was also a good deal of vague interpretations and understanding was lower for consumers with lower education.

As part of the CLYMBOL project, it was investigated whether adequate consumer understanding can be improved within the boundaries of the Regulation (Stancu et al., 2017). The regulation allows rewording of claims as long as the original meaning is retained, and it allows complementing of the specific health claim with a more general claim, as long as the specific claim is present as well. Both measures could potentially improve consumer understanding, by avoiding technical language or by providing a broader context. However, results from the study using several health claim examples did not provide any evidence on improved understanding; rather, the results indicated that providing more and more easily accessible information can indeed have the opposite effect, as consumers find it easier to make inferences from the health claim, including inferences that are unfounded.

2.4.3 Use in purchase decisions

Evidence concerning whether health claims are used in food purchase decisions is mixed (Hersey et al., 2013). Results on food choice and (self-reported) purchase intention vary: while some studies find that products with health claims increase intent to purchase (Coleman et al., 2014; Hwang et al., 2016; Lin et al., 2015), other studies show that they do not (Chrysochou and Grunert, 2014; Lalor et al., 2011;

Orquin and Scholderer, 2015). The outcome of such studies often depends on the claim- and consumer-related variables described earlier. For example choice of foods carrying health claims has been shown to be influenced by the type of carrier (Shepherd et al., 2012), consumer belief in the claimed benefit (Aschemann-Witzel et al., 2013; Petrovici et al., 2010) as well as awareness of the meaning of functional food (Krystallis and Chrysochou, 2011). Furthermore, claim type has been shown to significantly affect perceived credibility and purchase intention, whereas health-related motive orientations (both implicit and explicit) can translate into intent to purchase (Bialkova et al., 2016) or not (Hoefkens and Verbeke, 2013). One of the strongest factors influencing willingness to buy is personal relevance of a claim (Dean et al., 2012; Shepherd et al., 2012).

There is a research showing that health claims appear to have a greater impact during the initial stage of product knowledge. Those less familiar with a product tend to be more influenced by a health claim on-pack, compared with regular consumers of that product (Aschemann-Witzel et al., 2013; Moon et al., 2011). Consequently, it has been argued that claims and other health-related information may influence decision-making at first buy, while sensory factors such as taste determine loyalty and repeated consumption (Carrillo et al., 2012).

The use of claims in purchase behaviour can further vary by countries or regions. For example studies on the Japanese market showed a strong correlation between health claims and food purchase. Health claims increased the probability of purchase by 6%, with willingness to buy stronger for less price-sensitive consumers and weaker for consumers seeking naturalness in food products (Hirogaki, 2012). When consumers were health conscious, willingness to pay was found to be even stronger (Hirogaki, 2013). In a European cross-country study, the likelihood of buying products with health claims differed highly by country: German and Finnish consumers showed a greater likelihood of buying products with health claims, compared with Italian consumers who were less likely to buy a product when it carried a claim (Shepherd et al., 2012).

Another factor impacting the role of claims in food choice may be the age of the consumer. In experimental food auctions, younger participants reacted more positively to foods with health claims due to less prior knowledge about them (Hellyer et al., 2012). A study on 8- to 12-year-old children even showed that general health claims caused children to make unhealthier choices but could be reduced through health education (Miller et al., 2011).

Research is moving towards the notion that health claims are mostly perceived as positive by specific target groups, that is those that are familiar with the ingredient/benefit claimed and to whom the claim is relevant, either for themselves or for a family member with a health condition. Other consumers may, in fact, show adverse reactions to health claims (Lähteenmäki et al., 2010; Van Buul and Brouns, 2015).

Drivers of health claim use have been studied as part of the CLYMBOL project. In a 10-country study, CLYMBOL measured European consumers' ability and motivation to process health claims and how these two factors affected reported use of health claims. Results were similar for all claims tested (i.e. heart, brain, digestion, immunity, bone and tooth health) with a slightly higher motivation to process bone-related claims and a slightly lower motivation towards digestion-related claims. Country differences showed that both motivation and ability to process the information were highest among Spanish consumers and lowest among Dutch consumers. In line with findings from literature, consumers with a stronger personal need for health-related information, a stronger motivation to engage in preventive health behaviour and those who seek more stimulation during shopping had a stronger motivation to process health claims. Conversely, subjective (self-assessed) knowledge about the healthiness of foods emerged as the key driver for consumers' ability to process health claims and symbols. Interestingly, participants with greater health claim-related knowledge tended to be more able but less motivated to process health claims.

Overall, motivation to process the information emerged as the key driver for health claim use. Ability to process the information impacted health claim use to a much smaller extent. It was, however, strongly and positively influenced by motivation to process the information contained in health claims. Additionally, perceived credibility of health claims and attitude towards food carrying such claims was correlated with motivation and ability which in turn correlated with purchase intention for foods with health claims (Hung et al., 2017).

The overriding importance of motivation was corroborated in a series of experimental studies also carried out as part of the CLYMBOL project. Across countries, settings and types of health claims, a general result was that without a specific mindset or health goal, consumers do not usually choose food products with health claims (CLYMBOL, 2016). Having a health goal (which can be induced by priming) increases the likelihood of choosing products with health claims, and having a specific health goal (e.g. healthy bones) increases the likelihood of choosing products that carry a claim related to that specific goal. While this does not appear too surprising, it puts the effects of health claims on consumers into perspective: health claims have the highest chance of being used in consumers' food choices if they appeal to a specific niche with a specific health goal, not to the general population. This means that health claims addressing specific health benefits on products specifically developed to appeal to people with this health concern have the highest chance of being used. In line with previous research (Chrysochou and Grunert, 2014), the CLYMBOL studies also confirmed that implicit packaging cues such as images and colours can be as influential in determining how healthy a product is as a health claim.

2.4.4 Effects on consumption

Can health claims affect the quantity of a product consumed? It has been hypothesised that when a product is perceived as healthier, this may induce people to eat more because of a licensing effect, and there is indeed some evidence that nutrition claims such as 'low fat' can lead people to consume more (Belei et al., 2012; Wansink and Chandon, 2006), although the evidence is not unequivocal (Gravel et al., 2012). However, experimental studies conducted as part of the CLYMBOL project provided no evidence that this effect would extend to health claims as well.

2.5 A realistic perspective on the effect of health claims on consumers

Health claims have attracted enormous interest both from the food industry and from food regulators. The industry viewed products with health claims as a way towards growth with value-added products in otherwise mature markets, and regulators have seen health claims both as a measure to further inform consumer choices and as a source of concern because of their potential to mislead consumers. Both the enthusiasm and the concern are linked to how consumers react to health claims on food products: from an industry and from a regulatory perspective, the role of health claims hinges on how consumers perceive them and how this affects their food choices and dietary intake.

The research findings presented in this chapter suggest that both the industry and regulators may have overestimated the importance that consumers attach to health claims. The overarching result is that consumers are not terribly interested in health claims. Health claims are of relevance to consumers in their food choices only when consumers have a health motive when making the choice, and even then the perception of healthiness of a food product will be influenced by many other factors in addition to the health claim. This finding is not really surprising. Health claims address specific health benefits caused by specific functional ingredients. For the health claim to be relevant to consumers, consumers must not only have a health motive when making the choice, but the choice must also be related to the specific health benefit addressed by the claim. There is not much evidence suggesting that consumers use health claims as a key indicator when judging the overall healthfulness of a food product; in some cases even the opposite has been shown to take place (Lähteenmäki et al., 2010).

The role of health claims is therefore strongest in the context of products developed for specific target groups having specific health concerns. For such targets groups, both the general health interest and the specific interest in the health benefit addressed by the claim will be higher than that for the general population. On this motivational basis, the health claim can be part of consumers' food choice, and in these cases it is also appropriate to ask questions about adequate understanding and how it can be improved, how health claims interact with their context in bringing about perceptions of healthfulness and how they fit into people's overall diet.

References

Andrews, J.C., Netemeyer, R.G., Burton, S., 1998. Consumer generalization of nutrient content claims in advertising. J. Market. 62 (4), 62–75.

Aschemann-Witzel, J., 2011. The EU health claim regulation in international comparison: review of the possible impact on food marketing and consumer protection. CAB Rev. 6, 1–6.

Aschemann-Witzel, J., Maroschek, N., Hamm, U., 2013. Are organic consumers preferring or avoiding foods with nutrition and health claims? Food Qual. Preference 30, 68–76.

Asp, N.G., Bryngelsson, S., 2008. Health claims in Europe: new legislation and PASSCLAIM for substantiation. J. Nutr. 138, 1210S–1215S.

Bech-Larsen, T., Scholderer, J., 2007. Functional foods in Europe: consumer research, market experiences and regulatory aspects. Trends Food Sci. Technol. 18, 231–234.

Belei, N., Geyskens, K., Goukens, C., Ramanathan, S., Lemmink, J., 2012. The best of both worlds? Effects of attribute-induced goal conflict on consumption of healthful indulgences. J Market. Res. 49, 900–909.

Bialkova, S., Sasse, L., Fenko, A., 2016. The role of nutrition labels and advertising claims in altering consumers' evaluation and choice. Appetite 96, 38–46.

Bilman, E.M., van Kleef, E., Mela, D.J., Hulshof, T., van Trijp, H.C.M., 2012. Consumer understanding, interpretation and perceived levels of personal responsibility in relation to satiety-related claims. Appetite 59, 912–920.

Binns, N., 2014. The regulation of health claims in Europe. Sadler, M.J. (Ed.), Foods, Nutrients and Food Ingredients with Authorised EU Health Claims, vol. 1, Woodhead Publishing, Cambridge, United Kingdom, pp. 3–22, (Chapter 1).

Brookes, G., 2010. Economic Impact Assessment of the European Union (EU)'s Nutrition & Health Claims Regulation on the EU Food Supplement Sector and Market. UK Green Building Council, London. For the European Health Claims Alliance. Available from: www.pgeconomics.co.uk/pdf/Impact-Assessment-health-claims.pdf.

Bröring, S., Khedkar, S., Ciliberti, S., 2016. Reviewing the Nutrition and Health Claims Regulation (EC) No. 1924/2006: what do we know about its challenges and potential impact on innovation? Int. J. Food Sci. Nutr. 68, 1–9.

Buttriss, J., 2015. Nutrition and health claims in practice. Nutr. Bull. 40, 211–222.

Buttriss, J.L., Benelam, B., 2010. Nutrition and health claims: the role of food composition data. Eur. J. Clin. Nutr. 64, 8–13.

Carrillo, E., Varela, P., Fiszman, S., 2012. Effects of food package information and sensory characteristics on the perception of healthiness and the acceptability of enriched biscuits. Food Res. Int. 48, 209–216.

Cavaliere, A., De Marchi, E., Banterle, A., 2016. Does consumer health-orientation affect the use of nutrition facts panel and claims? An empirical analysis in Italy. Food Qual. Prefer. 54, 110–116.

Chrysochou, P., Grunert, K.G., 2014. Health-related ad information and health motivation effects on product evaluations. J. Bus. Res. 67, 1209–1217.

CLYMBOL, 2016. Deliverable 4.6: an integrated report on the totality of effects reported. Available from: www.clymbol.eu.

Coleman, K.L., Miah, E.M., Morris, G.A., Morris, C., 2014. Impact of health claims in prebiotic-enriched breads on purchase intent, emotional response and product liking. Int. J. Food Sci. Nutr. 65, 164–171.

Dean, M., Lampila, P., Shepherd, R., Arvola, A., Saba, A., Vassallo, M., Claupein, E., Winkelman, M., Lähteenmäki, L., 2012. Perceived relevance and foods with health-related claims. Food Qual. Prefer. 24, 129–135.

EU, 2006. Regulation (EC) No. 1924/2006 of the European Parliament and of the Council of 20 December 2006 on nutrition and health claims made on foods. Off. J. Eur. Union 30.12.2006, L 404.

EU, 2011. Regulation (EU) No. 1169/2011 of the European Parliament and of the Council of 25 October 2011 on the provision of food information to consumers, amending Regulations (EC) No. 1924/2006 and (EC) No. 1925/2006 of the European Parliament and of the Council, and repealing Commission Directive 87/250/EEC, Council Directive 90/496/EEC, Commission Directive 1999/10/EC, Directive 2000/13/EC of the European Parliament and of the Council, Commission Directives 2002/67/EC and 2008/5/EC and Commission Regulation (EC) No 608/2004. Off. J. Eur. Union. 22.11.2011, L 304/18.

European Parliament and Council of Europe, 2000. Directive 2000/13/EC of the European Parliament and of the Council of 20 March 2000 on the approximation of the laws of the Member States relating to the labelling, presentation and advertising of foodstuffs. Off. J. Eur. Commun. L 404, L 109/29.

European Parliament and Council of Europe, 2006. EC Regulation EC No. 1924/2006 of the European Parliament and of the Council of 20 December 2006 on nutrition and health claims made on foods. Off. J. Eur. Commun. L 404, L 12/3.

Gravel, K., Doucet, É., Herman, C.P., Pomerleau, S., Bourlaud, A.S., Provencher, V., 2012. Healthy," "diet," or "hedonic". How nutrition claims affect food-related perceptions and intake? Appetite 59, 877–884.

Grunert, K.G., Scholderer, J., Rogeaux, M., 2011. Determinants of consumer understanding of health claims. Appetite 56, 269–277.

Harris, J.L., Thompson, J.M., Schwartz, M.B., Brownell, K.D., 2011. Nutrition-related claims on children's cereals: what do they mean to parents and do they influence willingness to buy? Public Health Nutr. 1, 2207–2212.

Hellyer, N.E., Fraser, I., Haddock-Fraser, J., 2012. Food choice, health information and functional ingredients: an experimental auction employing bread. Food Policy 37, 232–245.

Hersey, J.C., Wohlgenant, K.C., Arsenault, J.E., Kosa, K.M., Muth, M.K., 2013. Effects of front-of-package and shelf nutrition labelling systems on consumers. Nutr. Rev. 71, 1–14.

Hieke, S., Kuljanic, N., Pravst, I., Miklavec, K., Kaur, A., Brown, K.A., Egan, B.M., Pfeifer, K., Gracia, A., Rayner, M., 2016. Prevalence of nutrition and health-related claims on prepackaged foods: a five-country study in Europe. Nutrients 8, 137.

Hieke, S., Kuljanic, N., Wills, J.M., Pravst, I., Kaur, A., Raats, M.M., van Trijp, H.C.M., Verbeke, W., Grunert, K.G., 2015. The role of health-related claims and health-related symbols in consumer behaviour: Design and conceptual framework of the CLYMBOL project and initial results. Nutr. Bull. 40, 66–72.

Hirogaki, M., 2012. How are functional foods perceived in Japan? Empirical study of young Japanese consumers' behaviour. Papers from the Ninth International CIRCLE Conference. Int. J. Manag. Cases, 185–199 (special issue).

Hirogaki, M., 2013. Estimating consumers' willingness to pay for health food claims: a conjoint analysis. Int. J. Innov. Manag. Technol. 4, 541–546.

Hoefkens, C., Verbeke, W., 2013. Consumers' health-related motive orientations and reactions to claims about dietary calcium. Nutrients 5, 82–96.

Hung, Y., Grunert, K.G., Hoefkens, C., Hieke, S., Verbeke, W., 2017. Motivation outweighs ability in explaining European consumers' use of health claims. Food Qual. Prefer. 58, 34–44.

Hwang, J., Lee, K., Lin, T.N., 2016. Ingredient labelling and health claims influencing consumer perceptions, purchase intentions, and willingness to pay. J. Foodserv. Bus. Res. 19, 352–367.

Kaur, A., Scarborough, P., Hieke, S., Kusar, A., Pravst, I., Raats, M., Rayner, M., 2016. The nutritional quality of foods carrying health-related claims in Germany, the Netherlands, Spain, Slovenia and the United Kingdom. Eur. J. Clin. Nutr. 70, 1388–1395.

Khedkar, S., Bröring, S., Ciliberti, S., 2016a. Exploring the Nutrition and Health Claims Regulation (EC) No. 1924/2006: what is the impact on innovation. Int. J. Food Sci. Nutr. 68 (1), 10–17.

Khedkar, S., Ciliberti, S., Bröring, S., 2016b. The EU health claims regulation: implications for innovation in the EU food sector. Br. Food J. 118, 2647–2665.

Krystallis, A., Chrysochou, P., 2011. Do health claims and prior awareness influence consumers' preferences for unhealthy foods? The case of functional children's snacks. Agribusiness 28, 86–102.

Lalor, F., Kennedy, J., Wall, P.G., 2011. Impact of nutrition knowledge on behaviour towards health claims on foodstuffs. Br. Food J. 113, 753–765.

Leathwood, P.D., Richardson, D.P., Strater, P., Todd, P.M., van Trijp, H.C.M., 2007. Consumer understanding of nutrition and health claims: sources of evidence. Br. J. Nutr. 98, 474–484.
Lin, Y.C., Lee, Yi, C., Wang, Y.F., 2015. Exploring the influence of tea beverage health claims on brand evaluation and purchase and intention. Int. J. Organ. Innov. 8, 88–99.
Lynam, A.M., McKevitt, A., Gibney, M.J., 2011. Irish consumers' use and perception of nutrition and health claims. Public Health Nutr. 14, 2213–2219.
Lähteenmäki, L., 2013. Claiming health in food products. Food Qual. Prefer. 27, 196–201.
Lähteenmäki, L., Lampila, P., Grunert, K.G., Boztug, Y., Ueland, Ø., Åström, A., Martinsdóttir, E., 2010. Impact of health-related claims on the perception of other product attributes. Food Policy 35, 230–239.
Miklavec, K., Pravst, I., Grunert, K.G., Klopcic, M., Pohar, J., 2015. The influence of health claims and nutritional composition on consumers' yoghurt preferences. Food Qual. Prefer. 43, 26–33.
Miller, E.G., Seiders, K., Kenny, M., Walsh, M.E., 2011. Children's use of on-package nutritional claim information. J. Consum. Behav. 10, 122–132.
Moon, W., Balasubramanian, S.K., Rimal, A., 2011. Health claims and consumers' behavioral intentions: the case of soy-based food. Food Policy 36, 480–489.
Orquin, Jacob L., Scholderer, Joachim, 2015. Consumer judgments of explicit and implied health claims on foods: misguided but not misled. Food Policy 51, 144–157.
Petrovici, D., Fearne, A., Nayga, Jr., R.M., Drolias, D., 2010. Nutritional knowledge, nutritional labels, and health claims on food. Br. Food J. 114, 768–783.
Raghunathan, R., Naylor, R.W., Hoyer, W.D., 2006. The unhealthy = tasty intuition and its effects on taste inferences, enjoyment, and choice of food products. J. Market. 70 (4), 170–184.
Richardson, D.P., 2012. Symposium 2: nutrition and health claims: help or hindrance preparing dossiers: strength of the evidence and problems of proof. Proc. Nutr. Soc. 71, 127–140.
Roe, B., Levy, A.S., Derby, B.M., 1999. The impact of health claims on consumer search and product evaluation outcomes: results from FDA experimental data. J. Public Policy Market. 18, 89–105.
Rogeaux, M., 2010. Consumer understanding and reaction to health claims. Insights and methodology. In: Jaeger, S.R., MacFie, H. (Eds.), Consumer-Driven Innovation in Food and Personal Care Products. Woodhead Publishing, Cambridge, United Kingdom, pp. 277–302.
Shepherd, R., Dean, M., Lampila, P., Arvola, A., Saba, A., Vassallo, M., Claupein, E., Winkelmann, M., Lähteenmäki, L., 2012. Communicating the benefits of wholegrain and functional grain products to European consumers. Trends Food Sci. Technol. 25, 63–69.
Stancu, V., 2016. Consumer understanding of health claims: motivation and ability related antecedents. In: Consumer Understanding of Health Claims: Methods for Assessment and Potential Antecedents, PhD dissertation. Available from: http://badm.au.dk/fileadmin/Business_Administration/PhD/PhD_dissertation_Violeta_Stancu.pdf.
Stancu, V., Grunert, K.G., Lähteenmäki, L., 2017. Consumer inferences from different versions of a beta-glucans health claim. Food Qual. Prefer. 60, 81–95.
Vadiveloo, M., Morwitz, V., Chandon, P., 2013. The interplay of health claims and taste importance on food consumption and self-reported satiety. Appetite 71, 349–356.
Van Buul, V.J., Brouns, F.J.P.H., 2015. Nutrition and health claims as marketing tools. Crit. Rev. Food Sci. Nutr. 55, 1552–1560.
Van Kleef, E., Van Trijp, H., Paeps, F., Fernandez-Celemin, L., 2008. Consumer preferences for front-of-pack calories labelling. Public Health Nutr. 11, 203–213.
van Trijp, H.C.M., van der Lans, I.A., 2007. Consumer perceptions of nutrition and health claims. Appetite 48, 305–324.
Van Wezemael, L., Caputo, V., Nayga, Jr., R.M., Chryssochoidis, G., Verbeke, W., 2014. European consumer preferences for beef with nutrition and health claims: a multi-country investigation using discrete choice experiments. Food Policy 44, 167–176.

Verhagen, H., van Loveren, H., 2016. Status of nutrition and health claims in Europe by mid 2015. Trends Food Sci. Technol. 56, 39–45.

Verhagen, H., Vos, E., Francl, S., Heinonen, M., van Loveren, H., 2010. Status of nutrition and health claims in Europe. Arch. Biochem. Biophys. 501, 6–15.

Wansink, B., Chandon, P., 2006. Can "low-fat" nutrition labels lead to obesity? J. Market. Res. 43, 605–617.

Wills, J.M., Storcksdieck Genannt Bonsmann, S., Kolka, M., Grunert, K.G., 2012. Symposium 2: nutrition and health claims: help or hindrance? European consumers and health claims: attitudes, understanding and purchasing behaviour. Proc. Nutr. Soc. 71, 229–236.

Wong, C.L., Arcand, J.A., Mendoza, J., Henson, S.J., Qi, Y., Lou, W., L'Abbé, M.R., 2013. Consumer attitudes and understanding of low-sodium claims on food: an analysis of health and hypertensive individuals. Am. J. Clin. Nutr. 97, 1288–1298.

Wong, C.L., Mendoza, J., Henson, S.J., Qi, Y., Lou, W., L'Abbe, M.R., 2014. Consumer attitudes and understanding of cholesterol-lowering claims on food: randomize mock-package experiments with plant sterol and oat fibre claims. Eur. J. Clin. Nutr 68, 946–952.

Further Reading

EC Register on nutrition and health claims made on food. Available from: http://ec.europa.eu/food/safety/labelling_nutrition/claims/register/public/?event=register.home.

Volkova, E., Mhurchu, C.N., 2015. The influence of nutrition labeling and point-of-purchase information on food behaviours. Curr. Obes. Rep. 4, 19–29.

Part Two

Authorised Reduction of Disease Risk Claims

Authorised EU health claims for calcium and calcium with vitamin D (for low bone mineral density and risk of fractures)

3

Philippa S. Gibson, Michele J. Sadler**, Susan A. Lanham-New**
*University of Surrey, Guildford, United Kingdom; **Rank Nutrition Ltd, Bethersden, Kent, United Kingdom

3.1 Introduction

3.1.1 Defining bone health

Bones break due to the loads placed on them exceeding the ability of the bone to absorb the energy involved. This may be as a result of a number of factors, including a reduction in bone mass, a change in the distribution of bone, loss of cancellous or cortical micro-architecture, an accumulation of damaged bone of a change in the material properties of the remaining bone. Throughout life, the skeleton requires optimum development and maintenance of its integrity since the resulting effect of poor bone health is osteoporotic fracture.

Osteoporosis is currently defined as a metabolic bone disease which has two predominant characteristics; low bone mass with micro-architectural deterioration of bone tissue. Both factors result in increased bone fragility and consequently increased risk of fracture. Bone 'weakness' refers to both poor structural quality and decreased bone mass.

3.1.2 Health economics of osteoporosis

Current figures estimate that one in two women and one in five men over the age of 50 years will experience a fracture as a result of low bone strength in their lifetime (Harvey et al., 2010). Approximately 300,000 fragility fractures occur in the United Kingdom alone each year (National Osteporosis Society, 2017), with medical costs predicted to be £6 billion by 2036 (NICE Clinical Guidelines, 2012). This is a significant cost contributing to the financial implementations of osteoporosis in Europe in 2010 was estimated at €37.4 billion (Hernlund et al., 2013), and is substantially higher than that of the USA ($22 billion) (Blume and Curtis, 2011). Hip fractures are believed to account for over a third of the total figure and are a reflection of the inpatient (hospital) and outpatient (nursing home) care required for patients presenting with osteoporosis. In the context of other conditions, osteoporosis is grossly underfunded in

Foods, Nutrients and Food Ingredients with Authorised EU Health Claims. http://dx.doi.org/10.1016/B978-0-08-100922-2.00003-6
Copyright © 2018 Elsevier Ltd. All rights reserved.

comparison to other diseases such as cardiovascular disease and cancer (Royal College of Physicians, 2000). Osteoporotic fractures are continually associated with both an increase in mortality and morbidity, giving rise to a huge burden on the health care system. In addition, there is tremendous pain and discomfort associated with osteoporosis for the patient. The lifetime risk for a fragility fracture in the United Kingdom is estimated at 53.2% for women and 20.7% in men aged 50 years and older (Dennison and Cooper, 2007).

Given that nearly 9 million fractures worldwide are associated with osteoporosis, and almost 65,000 UK people presenting with a hip fracture in 2015 (Boulton et al., 2016) set to rise to 101,000 in 2020 (Association, 2007), there can be no doubt that the future economic impact of osteoporosis will be phenomenal.

3.2 Evidence for effect of calcium alone on reduction of post-menopausal bone loss and reduced fracture risk

Calcium is an essential component of bone mineral, and therefore calcium is vital for skeletal health. The United Kingdom reference nutrient intake (RNI) for calcium in older men and women is 700 mg/day, aiming to prevent age related bone loss, while the World Health Organization/Food and Agriculture Organization of the United Nations (WHO/FAO) recommend 1000 mg/day for European adults. Current estimations for intakes based on the National Diet and Nutrition Survey is between 680 and 900 mg/day for women and men respectively living in the United Kingdom (Whitton et al., 2011), demonstrating lower intakes than the European RNI.

To date, clinical trials investigating the effect of increased calcium intake (either through dietary sources or supplementation) without other supplementation, such as vitamin D, on the reduction of post-menopausal bone loss and risk of bone fractures have been of relatively short duration, and it is difficult to determine if the positive effects of calcium supplementation on bone are maintained long-term. This is an area that requires urgent attention. Despite this, there are limited good data, which support calcium supplementation for the prevention of bone loss in post-menopausal women.

At present, only one long-term follow up study, spanning 10 years, investigating a 5-year randomised controlled trial (RCT), with placebo control, in healthy post-menopausal women aged >55 years living in New Zealand (n = 1471) demonstrated no effect of 1 g/day calcium citrate supplementation on total fracture and hip fracture 5 years after cessation of calcium supplementation. However, there were significant reductions in forearm and vertebral fractures in those assigned to calcium during the intervention (Radford et al., 2014). This study is suggestive that adequate intakes of calcium and pre-existing health status are likely to be important.

In the first of two recent systematic reviews and meta-analyses of RCTs, Tai et al. (2015) investigated extra dietary (15 studies; n = 1,533) or supplemental calcium (51 studies; n = 12,257) in women and men aged over 50, and if the effects on bone mineral density (BMD) were comparable. Tai et al. observed only a small increase in

BMD, with no further gains beyond the 1st year. Importantly, this limited improvement in bone density showed no greater effect when calcium was combined with vitamin D at any dose, even among participants with low serum concentration of 25-hydroxyvitamin D, a high calcium dose, or baseline dietary calcium intakes over 800 mg/day (Tai et al., 2015). In the second comprehensive systematic review, Bolland et al. identified 26 trials involving over 69,000 subjects, and whether increased calcium intake (from the diet and/or supplementation) could reduce fracture risk. While most of the reviewed trials tested supplementation, evidence of publication bias in small or moderate sized trials were identified. The meta-analyses confined to trials with the lowest risk of bias found that calcium supplements had no effect on risk of fractures at any site, and concluded that there was little evidence to support the theory that higher intakes of dietary calcium could reduce the risk of fractures (Bolland et al., 2015).

3.2.1 Dairy intakes as a source of dietary calcium

Although supplementation remains a good way of ensuring whether daily calcium requirements are met, dairy products are a naturally high source of bioavailable calcium. Additionally, milk and milk products are a source of other bone essential nutrients, including protein, magnesium, potassium and in some cases such as the United States of America, fortified with vitamin D. In the United Kingdom, current recommendations are 2–3 portions of dairy foods per day, with the aim of meeting the 700 mg calcium requirement for adults aged 19 years and older through dietary sources without the need for supplementation. Additionally, over supplementation of calcium has previously been shown to increase the risk of kidney stones and cardiovascular events, and therefore should be monitored carefully.

Work by Sahni et al. (2014) investigated the long-term effects of dairy consumption in 830 older aged, free-living men and women (mean age at baseline 77 years) taking part in the Framingham Original Cohort. A 20-year follow-up period demonstrated those who consumed medium (>1 and <7 servings per week) and high (≥7 servings per week) milk intakes tended to have a lower hip fracture risk in comparison to those with low (≤1 serving per week) milk intakes (Sahni et al., 2014), demonstrating the beneficial effects of dairy products in the diet for bone health. This effect was also noted in older aged women (80–92 years) where protein intakes from dairy sources were associated with appendicular bone mineralisation and muscle mass (Radavelli-Bagatini et al., 2014). However, in contradiction, an observational study conducted in two Swedish cohorts totalling over 10,000 subjects, indicated higher fracture risks in women consuming high quantities of milk as assessed by food frequency questionnaires. Results indicated a dose-dependant increase in fracture rates in women, but not men. Surprisingly, in a separate analysis, women who were high consumers of cheese and fermented dairy products had lower fracture risks than women who were low consumers; for every portion consumed, risk of fracture was reduced by 10%-15%. The authors conclude that due to the observational study design, cautious interpretation of the results should be advised (Michaelsson et al., 2014).

3.3 Effect of vitamin D alone on fall prevention: prevention of bone loss, fracture reduction, and muscle function and strength

Vitamin D is critical to bone health, through stimulation of the formation of bone matrix, bone maturation and increasing the uptake of calcium from the gut and kidneys. Vitamin D is also a key factor in post-menopausal bone loss and skeletal integrity in an ageing population. Guidelines for adults in the United Kingdom have been set at 400 IU/day (SACN, 2016), while the USA recommends 600 IU/day (Medicine, 2011).

3.3.1 Fall prevention

The effect of vitamin D on falls is discussed in the next chapter, in relation to the EU authorised reduction of disease risk health claim, whereby falls are considered a risk factor for bone fracture (Bischoff-Ferrari and Sadler, 2017). It is noteworthy that in five studies that had falls as a primary outcome, on which EFSA relied to substantiate the claim, vitamin D was used in combination with calcium, and the effect was compared with calcium alone. However, EFSA considered it unlikely that calcium is required for an effect of vitamin D on the risk of falling and thus concluded that the effect was established for vitamin D alone. However, the co-administration of calcium alongside vitamin D remains an important issue to consider (LeBlanc and Chou, 2015). EFSA also concluded that given the well-established role of vitamin D on muscle function, improvement in muscle function, strength and body balance offered a plausible mechanism by which vitamin D could exert the claimed effect, though recognising the relationship between vitamin D and muscle strength, physical performance and body balance in the elderly is yet to be established (EFSA, 2011).

3.3.2 Prevention of bone loss

Hansen et al. (2015) reported in a recent randomised, double-blind, placebo-controlled clinical trial conducted in post-menopausal women younger than 75 years with vitamin D insufficiency without osteoporosis found that neither low dose (800 IU/day) or high-dose (50,000 IU twice monthly) vitamin D treatment for 1 year significantly changed BMD; the mean levels of serum 25-hydroxyvitamin D in the placebo, low-dose and high-dose groups were 47, 70 and 140 nmol/L, respectively. Uusi-Rasi et al. conducted a 2-year randomised clinical trial that included over 400 home-dwelling women in Finland aged between 70 and 80 years and randomised to either: placebo without exercise, vitamin D (800 IU/day) without exercise, placebo and exercise, or vitamin D and exercise. Exercise consisted of supervised group training classes and the focus included balance, weights, agility and strengthening. In this study vitamin D helped to maintain bone density in the femoral neck and increased tibial trabecular density in the shinbone (Uusi-Rasi et al., 2015).

3.3.3 Fracture reduction

Supplementation on its own appears to have little, if any at all, effect on prevention of fractures in the elderly if the dosage is not adequate. In the most recent Cochrane review investigating vitamin D and vitamin D analogues for the prevention of fractures in post-menopausal women and older men (2014), 53 studies totalling 91,791 participants. Specifically looking at vitamin D alone, vitamin D appeared to be unlikely to be effective at reducing hip fracture or new fracture rates (Avenell et al., 2014).

3.3.4 Muscle function and strength

Alongside the conflicting evidence suggesting a role for vitamin D in bone strength and reduction of fracture risk is the role of vitamin D in muscle function and strength. Previous data from older-aged women is out of date, but demonstrated that higher fall rates were correlated with lower 25(OH)D serum concentrations (Flicker et al., 2003; Snijder et al., 2006), The most recent systematic review investigating the evidence of vitamin D, muscle function and strength demonstrated significant improvements in upper and lower limb strength when supplemented with vitamin D (Tomlinson et al., 2015). However, data were primarily investigated in younger females (67% female, age range 21.5–31.5 years) and therefore may not correlate with post-menopausal women and older-aged men. In the study by Uusi-Rasi et al. (2015) though neither vitamin D nor exercise reduced overall falls, exercise did improve muscle strength and balance. However, vitamin D did not enhance the effects of exercise on physical functioning, with conclusions that exercise in the form of strength and balance training had the most effect on fall rates (Uusi-Rasi et al., 2015). Data from Thailand investigating supplementation of post-menopausal women ($n = 52$) with 20,000 IU vitamin D_2 per week for 4 weeks showed significant improvements in balance compared with the control group (Anek et al., 2015), while treatment of Brazilian post-menopausal women ($n = 160$) with 1000 IU vitamin D_3 daily for 9 months resulted in improvements in postural balance and reduced falls (Cangussu et al., 2016) alongside reduced sarcopenia rates, control of loss of lean mass and improvements in muscles strength (Cangussu et al., 2015)

3.3.5 Vitamin D and bone health

As more high quality studies are conducted, including a promising study in over 5000 adults living in New Zealand set to publish data imminently (Scragg et al., 2016), reanalysis of the role that vitamin D plays in maintaining bone health will be required to further clarify the evidence as an intervention for fall rates. The RCT by Uusi-Rasi et al. (2015) and meta-analysis by Stubbs et al. (2015) reminds us that the strongest and most consistent evidence for prevention of serious falls is not a dietary intervention, but is for exercise which, in addition to many other health benefits that it provides, should remain one of the key focuses on healthy ageing of bones.

3.4 Calcium and vitamin D combined on osteoporosis prevention

Calcium and vitamin D in combination are critical to bone health, and are known to stimulate bone matrix formation and bone maturation, and have been shown to moderately reduce fracture rates, although the population-level intervention has yet to be confirmed as an effective public health strategy (Harvey et al., 2017). Current European guidelines for osteoporosis prevention and reduced fracture risk suggests including at least 1000 mg/day calcium and 800 IU/day vitamin D (Kanis et al., 2013).

In China, habitual calcium intakes can be low due to low consumption of dairy products as seen in Western countries. Estimated calcium intakes are between 288 and 948 mg/day based on 12 studies (Fang et al., 2016). In a North Chinese post-menopausal population ($n = 210$), supplementation of milk powder with different calcium concentrations was assessed over a period of 2 years. Participants were randomised to 300, 600 or 900 mg calcium per day as milk powder in conjunction with 800 IU vitamin D per day. The primary outcome measure was improvement in BMD as assessed by dual-energy X-ray absorptiometry (DEXA). Those patients randomised to the lower doses of calcium supplementation demonstrated significant decrease in BMD at the spine and greater trochanter, while increases were observed in the 600–900 mg/day groups at Ward's triangle. Therefore, for healthy post-menopausal women, calcium supplementation through milk powder is a possible method for preventing bone loss in the greater trochanter (Chen et al., 2016).

In the recent meta-analysis conducted by the National Osteoporosis Foundation (Weaver et al., 2016a,b), the role of calcium with vitamin D supplementation from RCTs was investigated in the prevention of fractures in a total of 970 adults. Results reporting 192 hip fractures and 2231 total fractures demonstrated a 14% reduction in the risk of total fractures and a 39% reduction in risk of a hip fracture. The authors concluded supporting the use of calcium with vitamin D supplementation as an intervention for reducing fracture risks in both community-dwelling and institutionalised middle-aged to older adults (Weaver et al., 2016a,b). Adding to this, the previously mentioned Cochrane review highlighted the evidence supporting the role of vitamin D plus calcium supplementation resulted in a small reduction in hip fractures. Additionally, in relatively low risk, free-living populations there was a reduced rate of hip fracture of 1 rate per 1000 older adults per year. This effect was further improved in high risk populations, such as residents in institution, supplementation of vitamin D with calcium resulted in an average of 9 fewer fractures per 1000 older adults per year (Avenell et al., 2014). This data for older aged persons living in institutionalised residences is also supported in a meta-analysis which identified a reduced risk of hip fracture when vitamin D and calcium were co-administered, but this data was only derived from 2 studies (Bolland et al., 2014), with the authors remaining doubtful about the effect of vitamin D and calcium supplementation in free-living older population on fracture risk in a shortly followed second meta-analysis (Reid, 2015). Furthermore, the meta-analysis by Guo et al. (2014) identified vitamin D and calcium supplementation (pooled results of 11 RCTs) in non-institutionalised older-aged persons without cognitive impairment had a positive effect on number of falls (Guo et al., 2014).

While a large percentage of clinical trials investigating calcium and vitamin D supplementation target post-menopausal women due to the increased risk of fractures, improvement on older male BMD is of equal importance. In one recent systematic review with meta-analysis, there was limited evidence to support the role of calcium and vitamin D supplementation for improvements in male BMD. However, the authors concluded that high quality RCTs were limited in male cohorts, and further investigations were needed to determine any preventative benefits of supplementation (Silk et al., 2015). The conclusion of the consensus meeting of the European Society for Clinical and Economic Aspects of Osteoporosis, Osteoarthritis and Musculoskeletal Diseases (ESCEO) and the International Foundation for Osteoporosis (IOF) were that calcium and vitamin D supplementation resulted in a modest reduction in fracture risk, but supplementation of calcium alone was not currently supported by literature. In addition, vitamin D supplementation, rather than calcium supplementation, may reduce the risk of falls (Harvey et al., 2017).

3.5 Current EU health claims for calcium, and calcium and vitamin D combined

Two Article 14.1(a) reduction of disease risk claims have been authorised for reduced loss of bone mineral in post-menopausal women, low BMD being a risk factor for osteoporotic bone fractures. One claim is for calcium alone, which can only be used for food providing at least 400 mg of calcium per quantified portion. This claim is specific to women aged 50 years and older, and beneficial effects are obtained with a daily intake of at least 1200 mg of calcium from all sources. The other claim is for calcium in conjunction with vitamin D and this is also targeted at post-menopausal women. However this claim can only be used on food supplements, not on foods, and to use the claim they must provide at least 400 mg calcium and 15 μg of vitamin D per daily intake. The beneficial effect is obtained with a daily intake of at least 1200 mg of calcium and 20 μg of vitamin D from all sources.

Other authorised health claims for calcium include the Article 13.1 general function claims that relate to the maintenance of normal bones and normal teeth. These can only be used for foods, which are at least a source of calcium (15% of nutrient reference value—i.e. 120 mg per 100 g or 100 mL or per package if the package contains only a single portion). Additional Article 13.1 health claims for calcium are related to cardiac function, muscle function, function of digestive enzymes and energy yielding metabolism. Two health claims targeted at children have also been authorised, one for calcium alone and one for calcium in conjunction with vitamin D (Table 3.1).

3.5.1 Background to the Article 14.1(a) reduction of disease risk claims

EFSA issued an opinion in 2009, concluding that there was sufficient evidence to demonstrate cause and effect for a reduction of disease risk claim relating to reduced loss of bone mineral in post-menopausal women as the risk factor for development

Table 3.1 **Authorised EU health claims for calcium, and calcium with vitamin D**

Health claim types	Nutrient	Authorised claim wording	Conditions/restrictions of use
14.1(a) Reduction of disease risk claim	Calcium	Calcium helps to reduce the loss of bone mineral in post-menopausal women. Low BMD is a risk factor for osteoporotic bone fractures	The claim may be used only for food, which provides at least 400 mg calcium per quantified portion. Consumer to be informed that the claim is intended for women 50 years and older and that the beneficial effect is obtained with a daily intake of at least 1200 mg calcium from all sources. For foods with added calcium the claim may be used only for those targeting women 50 years and older
14.1(a) Reduction of disease risk claim	Calcium and vitamin D	Calcium and vitamin D help to reduce the loss of bone mineral in post-menopausal women. Low BMD is a risk factor for osteoporotic bone fractures	The claim may be used only for food supplements, which provide at least 400 mg calcium and 15 μg vitamin D per daily intake. Consumer to be informed that the claim is intended for women 50 years and older and the beneficial effect is obtained with a daily intake of at least 1200 mg calcium and 20 μg vitamin D from all sources. The claim may be used only for food supplements targeting women 50 years and older
14.1(b) Children's claim	Calcium	Calcium is needed for normal growth and development of bone in children	The claim may be used only for food which is at least a source of calcium as referred to in the claim SOURCE OF [NAME OF VITAMIN/S] AND/OR [NAME OF MINERAL/S] as listed in the Annex to Regulation (EC) No. 1924/2006—that is 15% of the Nutrient Reference Value (120 mg/100 g or 100 mL or package if the package contains only a single portion)

14.1(b) Children's claim	Calcium and vitamin D	Calcium and vitamin D are needed for normal growth and development of bone in children	As above
Art.13(1) General function claim	Calcium	Calcium contributes to normal blood clotting	As above
Art.13(1) General function claim	Calcium	Calcium contributes to normal energy-yielding metabolism	As above
Art.13(1) General function claim	Calcium	Calcium contributes to normal muscle function	As above
Art.13(1) General function claim	Calcium	Calcium contributes to normal neurotransmission	As above
Art.13(1) General function claim	Calcium	Calcium contributes to the normal function of digestive enzymes	As above
Art.13(1) General function claim	Calcium	Calcium has a role in the process of cell division and specialisation	As above
Art.13(1) General function claim	Calcium	Calcium is needed for the maintenance of normal bones	As above
Art.13(1) General function claim	Calcium	Calcium is needed for the maintenance of normal teeth	As above

BMD, Bone mineral density.

of osteoporotic bone fractures (EFSA, 2009). This was based on evidence published prior to 2009. Though the application was for calcium and vitamin D supplements, EFSA's opinion applied to all forms of calcium and vitamin D naturally occurring in foods, to forms authorised for addition to foods, and for forms authorised for use in food supplements. EFSA also concluded that the evidence consistently supported a cause and effect relationship for calcium alone, and for the combined intake of calcium and vitamin D. However, on the basis of the evidence submitted to substantiate the claim, EFSA was unable to set conditions of use for both foods and supplementary sources of calcium and vitamin D. The EU Commission thus tasked EFSA to consider this issue further. In 2010 EFSA published a further opinion on which the conditions of use could be based (EFSA, 2010). EFSA proposed that at least 1200 mg calcium from all sources or at least 1200 mg calcium with 800 IU of vitamin D from all sources needed to be consumed daily to achieve the health benefit, and that the conditions of use should be based on these amounts. As the daily intakes to produce the beneficial effect were higher than usual dietary recommendations, the EU Commission further tasked EFSA to review the upper safe levels (ULs) for calcium and vitamin D. These opinions were issued in 2012 (EFSA, 2012a,b). EFSA maintained the UL for calcium at the previous level or 2500 mg/day, as no new data was available. However, EFSA revised the UL for vitamin D from the previous level of 50 μg/day, setting a new UL of 100 μg/day for adults. It was thus considered that the proposed conditions of use for the two reduction of disease risk claims were within safe levels of intake, although it is important to note that 1200 mg for calcium is much higher than the current UK RNI for calcium (700 mg/day for adults). The claims were eventually authorised in November 2014, taking 5 years since the initial EFSA opinion was published.

3.6 Areas for future research

At present, data suggest a role for calcium in maintaining bone health with particular respect to BMD. Systematic reviews have demonstrated that calcium supplementation or improved dietary calcium and reduced fracture risks at several sites across the body. However, these improvements in BMD is often small, around 1%-2%, and therefore not clinically significant for reduction of fracture risk alone. However, BMD is only a surrogate marker for clinical risk of fracture. Further studies should investigate the source of calcium, with specific reference to dairy sources as acid-base or protein-calcium interactions may have a potential impact on risk of fracture, as current data is conflicting.

References

Anek, A., Bunyaratavej, N., Jittivilai, T., 2015. Effects of short-term vitamin D supplementation on musculoskeletal and body balance for prevention of falling in postmenopausal women. J. Med. Assoc. Thai. 98, S26–S31.

Darowski, A., 2007. The Care of Patients With Fragility Fracture. British Orthopaedic Association, London.

Avenell, A., Mak, J.C., O'connell, D., 2014. Vitamin D and vitamin D analogues for preventing fractures in post-menopausal women and older men. Cochrane Database Syst. Rev. 14 (4), CD000227.

Bischoff-Ferrari, H.A., Sadler, M.J., 2017. Authorised EU health claim for vitamin D and reduced risk of falls. In: Sadler, M.J. (Ed.), Foods, Nutrients and Food Ingredients with Authorised EU Health Claims, 3, Woodhead Publishing, Cambridge, (Chapter 4).

Blume, S.W., Curtis, J.R., 2011. Medical costs of osteoporosis in the eldery Medicare population. Osteoporos. Int. 22 (6), 1835–1844.

Bolland, M.J., Grey, A., Gamble, G.D., Reid, I.R., 2014. The effect of vitamin D supplementation on skeletal, vascular, or cancer outcomes: a trial sequential meta-analysis. Lancet Diab. Endocrinol. 2, 307–320.

Bolland, M.J., Leung, W., Tai, V., Bastin, S., Gamble, G.D., Grey, A., Reid, I.R., 2015. Calcium intake and risk of fracture: systematic review. BMJ 351, h4580.

Boulton, C., Bunning, T., Burgon, V., Cromwell, D., Johansen, A., Rai, S., Stanley, R., Tsang, C., Wakerman, R., 2016. National Hip Fracture Database Annual Report 2016.

Cangussu, L.M., Nahas-Neto, J., Orsatti, C.L., Bueloni-Dias, F.N., Nahas, E.A., 2015. Effect of vitamin D supplementation alone on muscle function in postmenopausal women: a randomized, double-blind, placebo-controlled clinical trial. Osteoporos Int. 26, 2413–2421.

Cangussu, L.M., Nahas-Neto, J., Orsatti, C.L., Poloni, P.F., Schmitt, E.B., Almeida-Filho, B., Nahas, E.A., 2016. Effect of isolated vitamin D supplementation on the rate of falls and postural balance in postmenopausal women fallers: a randomized, double-blind, placebo-controlled trial. Menopause 23, 267–274.

Chen, Y., Xiao, Y., Xie, B., Zhang, Q., MA, X., LI, N., Liu, M., Zhang, Q., 2016. Effect of milk powder supplementation with different calcium contents on bone mineral density of postmenopausal women in northern China: a randomized controlled double-blind trial. Calcif. Tissue Int. 98, 60–66.

Dennison, E., Cooper, C., 2007. Epidemiology including fragility fracture. In: Clunie, G., Keen, R. (Eds.), Osteoporosis (Oxford Rheumatology Library). Oxford University Press, Oxford.

EFSA, 2009. Scientific substantiation of a health claim related to calcium plus vitamin D3 chewing tablets and reduction of the risk of osteoporotic fractures by reducing bone loss pursuant to Article 14 of Regulation (EC) No 1924/2006. EFSA J. 1180, 1–13.

EFSA, 2010. Scientific opinion in relation to the authorisation procedure for health claims on calcium and vitamin D and the reduction of the risk of osteoporotic fractures by reducing bone loss pursuant to Article 14 of Regulation (EC) No 1924/20061. EFSA J. 8 (5), 1609.

EFSA, 2011. Scientific opinion on the substantiation of a health claim related to vitamin D and risk of falling pursuant to Article 14 of Regulation (EC) No 1924/2006. Scientific opinion of the panel on dietetic products, nutrition and allergies. EFSA J. 9 (9), 2382.

EFSA, 2012a. Scientific opinion on the tolerable upper intake level of calcium. EFSA J. 10 (7), 2814.

EFSA, 2012b. Scientific opinion on the tolerable upper intake level of vitamin D. EFSA J. 10 (7), 2813.

Fang, A.P., LI, K.J., Shi, H.Y., HE, J.J., LI, H., 2016. Habitual dietary calcium intakes and calcium metabolism in healthy adults Chinese: a systematic review and meta-analysis. Asia Pac. J. Clin. Nutr. 25, 776–784.

Flicker, L., Mead, K., Macinnis, R.J., Nowson, C., Scherer, S., Stein, M.S., Thomasx, J., Hopper, J.L., Wark, J.D., 2003. Serum vitamin D and falls in older women in residential care in Australia. J. Am. Geriatr. Soc. 51, 1533–1538.

Guo, J.L., Tsai, Y.Y., Liao, J.Y., TU, H.M., Huang, C.M., 2014. Interventions to reduce the number of falls among older adults with/without cognitive impairment: an exploratory meta-analysis. Int. J. Geriatr. Psychiatry 29, 661–669.

Hansen, K.E., Johnson, R.E., Chambers, K.R., Johnson, M.G., Lemon, C.C., VO, T.N., Marvdashti, S., 2015. Treatment of vitamin D insufficiency in postmenopausal women: a randomized clinical trial. JAMA Intern. Med. 175, 1612–1621.

Harvey, N.C., Biver, E., Kaufman, J.M., Bauer, J., Branco, J., Brandi, M.L., Bruyere, O., Coxam, V., Cruz-Jentoft, A., Czerwinski, E., Dimai, H., Fardellone, P., Landi, F., Reginster, J.Y., Dawson-Hughes, B., Kanis, J.A., Rizzoli, R., Cooper, C., 2017. The role of calcium supplementation in healthy musculoskeletal ageing : an expert consensus meeting of the European Society for Clinical and Economic Aspects of Osteoporosis, Osteoarthritis and Musculoskeletal Diseases (ESCEO) and the International Foundation for Osteoporosis (IOF). Osteoporos. Int. 28 (2), 447–462.

Harvey, N., Dennison, E., Cooper, C., 2010. Osteoporosis: impact on health and economics. Nat. Rev. Rheumatol. 6, 99–105.

Hernlund, E., Svedbom, A., Ivergard, M., Compston, J., Cooper, C., Stenmark, J., Mccloskey, E.V., Jonsson, B., Kanis, J.A., 2013. Osteoporosis in the European Union: medical management, epidemiology and economic burden. A report prepared in collaboration with the International Osteoporosis Foundation (IOF) and the European Federation of Pharmaceutical Industry Associations (EFPIA). Arch. Osteoporos. 8, 136.

Kanis, J.A., Mccloskey, E.V., Johansson, H., Cooper, C., Rizzoli, R., Reginster, J.Y., Scientific Advisory Board of the European Society for Clinical, Economic Aspects of Osteoporosis and Osteoarthritis, the Committee of Scientific Advisors of the International Osteoporosis Foundation, 2013. European guidance for the diagnosis and management of osteoporosis in postmenopausal women. Osteoporos. Int. 24, 23–57.

LeBlanc, E.S., Chou, R., 2015. Vitamin D and falls-fitting new data with current guidelines. JAMA Intern. Med. 175, 712–713.

Medicine, I.O., 2011. Dietary Reference Intakes for Calcium and Vitamin D. The National Academies Press, Washington, DC.

Michaelsson, K., Wolk, A., Langenskiold, S., Basu, S., Warensjo Lemming, E., Melhus, H., Byberg, L., 2014. Milk intake and risk of mortality and fractures in women and men: cohort studies. BMJ 349, g6015.

National Osteoporosis Society, 2017. British Orthopaedic Association 2007. Available from: http://www.bgs.org.uk/pdf_cms/pubs/Blue%20Book%20on%20fragility%20fracture%20care.pdf.

NICE Clinical Guidelines, 2012. Osteoporosis: Fragility Fracture Risk Costing Report: Implementing NICE Guidance. Available from: https://www.nice.org.uk/guidance/cg146/resources/costing-report-pdf-186811885.

Radavelli-Bagatini, S., Zhu, K., Lewis, J.R., Prince, R.L., 2014. Dairy food intake, peripheral bone structure, and muscle mass in elderly ambulatory women. J. Bone Miner. Res. 29, 1691–1700.

Radford, L.T., Bolland, M.J., Mason, B., Horne, A., Gamble, G.D., Grey, A., Reid, I.R., 2014. The Auckland calcium study: 5-year post-trial follow-up. Osteoporos. Int. 25, 297–304.

Reid, I.R., 2015. Effects of vitamin D supplements on bone density. J. Endocrinol. Invest. 38, 91–94.

Royal College of Physicians, 2000. Osteoporosis Clinical Guidelines for Prevention and Treatment. Royal College of Physicians, London.

SACN, 2016. Vitamin D and Health. The Stationary Office, London.

Sahni, S., Mangano, K.M., Tucker, K.L., Kiel, D.P., Casey, V.A., Hannan, M.T., 2014. Protective association of milk intake on the risk of hip fracture: results from the Framingham Original Cohort. J. Bone Miner. Res. 29, 1756–1762.

Scragg, R., Waayer, D., Stewart, A.W., Lawes, C.M., Toop, L., Murphy, J., Khaw, K.T., Camargo, JR, C.A., 2016. The Vitamin D Assessment (ViDA) Study: design of a randomized controlled trial of vitamin D supplementation for the prevention of cardiovascular disease, acute respiratory infection, falls and non-vertebral fractures. J. Steroid Biochem. Mol. Biol. 164, 318–325.

Silk, L.N., Greene, D.A., Baker, M.K., 2015. The effect of calcium or calcium and vitamin D Supplementation on bone mineral density in healthy males: a systematic review and meta-analysis. Int. J. Sport Nutr. Exerc. Metab. 25, 510–524.

Snijder, M.B., Van Schoor, N.M., Pluijm, S.M., Van Dam, R.M., Visser, M., Lips, P., 2006. Vitamin D status in relation to one-year risk of recurrent falling in older men and women. J. Clin. Endocrinol. Metab. 91, 2980–2985.

Stubbs, B., Brefka, S., Denkinger, M.D., 2015. What works to prevent falls in community-dwelling older adults? Umbrella review of meta-analyses of randomized controlled trials. Phys. Ther. 95, 1095–1110.

Tai, V., Leung, W., Grey, A., Reid, I.R., Bolland, M.J., 2015. Calcium intake and bone mineral density: systematic review and meta-analysis. BMJ 351, h4183.

Tomlinson, P.B., Joseph, C., Angioi, M., 2015. Effects of vitamin D supplementation on upper and lower body muscle strength levels in healthy individuals. A systematic review with meta-analysis. J. Sci. Med. Sport 18, 575–580.

Uusi-Rasi, K., Patil, R., Karinkanta, S., Kannus, P., Tokola, K., Lamberg-Allardt, C., Sievanen, H., 2015. Exercise and vitamin D in fall prevention among older women: a randomized clinical trial. JAMA Intern. Med. 175, 703–711.

Weaver, C.M., Alexander, D.D., Boushey, C.J., Dawson-Hughes, B., Lappe, J.M., Leboff, M.S., Liu, S., Looker, A.C., Wallace, T.C., Wang, D.D., 2016a. Calcium plus vitamin D supplementation and risk of fractures: an updated meta-analysis from the National Osteoporosis Foundation. Osteoporos. Int. 27, 367–376.

Weaver, C.M., Dawson-Hughes, B., Lappe, J.M., Wallace, T.C., 2016b. Erratum and additional analyses re: calcium plus vitamin D supplementation and the risk of fractures: an updated meta-analysis from the National Osteoporosis Foundation. Osteoporos. Int. 27, 2643–2646.

Whitton, C., Nicholson, S.K., Roberts, C., Prynne, C.J., Pot, G.K., Olson, A., Fitt, E., Cole, D., Teucher, B., Bates, B., Henderson, H., Pigott, S., Deverill, C., Swan, G., Stephen, A.M., 2011. National Diet and Nutrition Survey: UK food consumption and nutrient intakes from the first year of the rolling programme and comparisons with previous surveys. Br. J. Nutr. 106, 1899–1914.

Further Reading

Burge, R.T., Worley, D., Johansen, A., Bhattacharyya, S., Bose, U., 2001. The cost of osteoporotic fractures in the UK: projections for 2000–2020. J. Med. Econ. 4, 51–62.

Authorised EU health claim for Vitamin D and reduced risk of falls

4

Heike A. Bischoff-Ferrari, **, Michele J. Sadler*[†]
*University Hospital and University of Zurich, Zurich, Switzerland;
**University Clinic for Acute Geriatric Care, City Hospital Waid, Zurich, Switzerland;
[†]Rank Nutrition Ltd, Bethersden, Kent, United Kingdom

4.1 Introduction

This chapter will summarise the authorised EU health claim for vitamin D, reduced risk of falls and fracture prevention in older people, particularly focusing on the mechanistic and clinical evidence that links vitamin D to muscle health and fall prevention. Evidence from randomised-controlled trials (RCTs) for fall prevention will be reviewed based on the most recent clinical trials and meta-analyses of clinical trials, including those published since publication of the EFSA opinion for the claim (EFSA, 2009). Further, fall prevention efficacy will be explored by dose of vitamin D, 25-hydroxyvitamin D status, type of dwelling and treatment interval. Finally, this chapter will explore if fall prevention with vitamin D translated into fracture prevention among seniors at risk of vitamin D deficiency. To understand public-health relevance of vitamin D treatment for fall prevention, the epidemiology and cost of falls will be reviewed at the beginning of the chapter.

4.2 Epidemiology of falls and cost of falls

Each year 1 out of 3 persons age 65 years and older, and 1 out of 2 age 80 years and older experience at least one fall (Tinetti, 1994). Nine percent of falls require an emergency room visit (Bischoff et al., 2003), and serious injuries occur with 10%–15% of falls, resulting in fractures in 5% and hip fracture in 1%–2% (CDC, 2006). Additionally, falls are an independent determinant of functional decline and 40% of all nursing home admissions are due to a fall (Tinetti and Williams, 1997). Moreover, the primary risk factor for a hip fracture is a fall, and over 90% of all fractures occur after falling (Cummings et al., 1995). Recurrent fallers may have close to a fourfold-increased odds of sustaining a fall-related fracture, compared with individuals having a single fall (Pluijm et al., 2006). As the number of seniors aged 65 years and older is predicted to increase from 25% to 40% by 2030 (Eberstadt and Groth, 2007), the number of fall-related fractures would increase substantially. Notably, even today 75% of fractures

Foods, Nutrients and Food Ingredients with Authorised EU Health Claims. http://dx.doi.org/10.1016/B978-0-08-100922-2.00004-8
Copyright © 2018 Elsevier Ltd. All rights reserved.

occur among seniors aged 65 years and older (Melton et al., 1999), and by 2050 the worldwide incidence in hip fractures is expected to increase by 240% among women and 310% among men (Gullberg et al., 1997). Because of the increasing proportion of older individuals, annual costs from all fall-related injuries in the United States, in persons 65 years or older, were projected to increase from $20.3 billion in 1994 to $32.4 billion in 2020, including medical, rehabilitation, hospital costs and the costs of morbidity and mortality (Englander et al., 1996). In fact, fall injuries are among the 20 most expensive medical conditions amounting to $34 billion annually of direct medical costs for fall injuries (Stevens et al., 2006). Thus, therapeutic interventions in fall prevention that are effective, well tolerated and affordable at a broad public health level, are urgently needed. Vitamin D may have such a profile and will be reviewed in this chapter.

4.3 Mechanistic evidence on how vitamin D is related to muscle health

Mechanistically, four lines of evidence support a link between vitamin D and muscle health. *First*, the vitamin D receptor (VDR) is expressed in human muscle tissue as suggested in all but one study (Bischoff-Ferrari, 2012). *Second*, the vitamin D bound to its nuclear receptor in muscle tissue may lead to de novo protein synthesis (Ceglia et al., 2013), followed by a relative increase in the diameter and number of fast type II muscle fibres (Ceglia et al., 2013). In fact, supplementation with 4000 IU vitamin D compared with placebo increased the number of VDRs in muscle tissue, and the number and diameter of type II muscle fibres, among postmenopausal women and compared with placebo (Ceglia et al., 2013). At a clinical level this is highly relevant as fast type II muscle fibres decline with age relative to slow type I muscle fibres, and the decline in fast type II muscle fibres with age likely contributes to the increased propensity to fall among senior adults. *Third*, the enzyme 1-alpha-hydroxylase which transforms 25-hydroxyvitamin D into its active form 1,25-dihydroxyvitamin D is present in muscle tissue (Srikuea et al., 2012). *Fourth*, mice lacking the VDR show a skeletal muscle phenotype with smaller and variable muscle fibres and persistence of immature muscle gene expression during adult life (Bouillon et al., 2008; Endo et al., 2003). These abnormalities persist after correction of systemic calcium metabolism by a rescue diet (Endo et al., 2003) and suggest that vitamin D is relevant to muscle development at young age and maintenance of muscle health in later life.

4.4 Clinical evidence on how vitamin D is related to muscle health

Four lines of clinical evidence support a role of vitamin D in muscle health (Bischoff-Ferrari, 2012). *First*, proximal muscle weakness is a prominent feature of the clinical syndrome of vitamin D deficiency (Al-Shoha et al., 2009). *Second*, clinical findings in vitamin D deficiency myopathy include proximal muscle weakness, diffuse muscle pain and gait impairments such as a 'waddling' way of walking (Schott and Wills, 1976). Muscle weakness is an important risk factor for falls and is a prominent feature of the clinical syndrome of vitamin D deficiency (Schott and Wills, 1976). Thus, muscle weakness due to vitamin D deficiency may plausibly mediate fracture risk through an increased susceptibility to falls. *Third*, several observational studies suggest a positive association between 25-hydroxyvitamin D and muscle strength, or lower extremity function in older persons (Bischoff-Ferrari et al., 2004d; Wicherts et al., 2007). These findings were confirmed by several recent double-blind RCTs with 800 IU vitamin D3 resulting in a 4%–11% gain in lower extremity strength or function (Pfeifer et al., 2009; Bischoff et al., 2003), and up to 28% improvement in body sway (Pfeifer et al., 2000, 2009) in older adults age 65+ years at risk of vitamin D deficiency and treated with vitamin D supplements between 2 and 12 months (Muir and Montero-Odasso, 2011). Notably, however, such a benefit may be seen preferentially among senior adults with vitamin D deficiency, while vitamin D replete older adults may not benefit in the same way (Uusi-Rasi et al., 2015), as supported by a recent meta-analysis on vitamin D treatment and functional improvement (Stockton et al., 2011).

Fourth, vitamin D supplementation reduced the risk of falling in community-dwelling (Bischoff-Ferrari et al., 2006; Pfeifer et al., 2000, 2009) as well as in institutionalised senior adults (Bischoff et al., 2003; Broe et al., 2007) at risk of vitamin D deficiency in several double-blind RCTs, summarised in a 2009 meta-analysis discussed later (Bischoff-Ferrari et al., 2009a) and in other meta-analyses published on the topic and summarised in Table 4.1. Finally, it is important to note that vitamin D may address several components of the fall-fracture construct, including strength (Bischoff et al., 2003), balance (Pfeifer et al., 2000), lower extremity function (Cangussu et al., 2016), falling (Bischoff-Ferrari et al., 2009a), bone density (Bischoff-Ferrari et al., 2004c; Dawson-Hughes et al., 1997), the risk of hip and non-vertebral fractures (Bischoff-Ferrari et al., 2009b, 2012), and the risk of nursing home admission (Visser et al., 2006).

4.5 Vitamin D and Type II fast twitch muscle fibres and relevance for fall risk

Type II muscle fibres are fast-twitch fibres and therefore are the first to be recruited when fast reaction is needed such as in the prevention of a fall. Notably, ageing itself has been associated with a decrease in Type II fast-twitch relative to Type I slow-twitch muscle fibres (Grimby and Saltin, 1983). Given the high prevalence of vitamin D deficiency in senior adults, it is possible that the age related decline in Type II muscle fibres is in part explained by vitamin D deficiency, which may be accompanied by a decrease in muscular VDR expression with age (Bischoff-Ferrari et al., 2004a). In fact, muscle biopsy studies in humans suggest a potentially selective effect of vitamin D on Type II muscle fibres. Patients with osteomalacic myopathy reveal Type II muscle atrophy in muscle histology investigations (Yoshikawa et al., 1979). Notably,

Table 4.1 Meta-analyses of clinical trials testing vitamin D on fall prevention

Year	Author; journal	(%) Fall reduction with vitamin D	Effect size 95% CI
2004	Bischoff-Ferrari et al., 2004b; JAMA	−22% with supplemental and active vitamin D *(only double-blind RCTs)*	OR = 0.78 (0.64–0.92)
2007	Jackson et al., 2007; QiM	−12% with vitamin D3 supplementation	RR = 0.88 (0.78–1.00)
2008	O'Donnell et al., 2008; J. Bone Miner. Metab.	−34% with active vitamin D	OR = 0.66 (0.44–0.98)
2008	Richy et al., 2008; Calcif. Tissue Int.	−21% with active vitamin D	RR = 0.79 (0.64–0.96)
2009	Bischoff-Ferrari et al., 2012; BMJ	−19% with 700–1000 IU/day +10% with 200–600 IU/day of vitamin D3/2 supplementation *(only double-blind RCTs)*	RR = 0.81 (0.71–0.92) RR = 1.10 (0.89–1.35)
2010	Kalyani et al., 2010; J. Am. Ger. Soc.	−14% with 200–100 IU/day	RR = 0.86 (0.79–0.93)
2010	Cameron et al., 2010; Cochrane Syst. Rev.	−28% with vitamin D supplementation	RR = 0.72 (0.55–0.95) *rate*
2011	Michael et al., 2011; Ann. Intern. Med.	−17% with vitamin D supplementation	RR = 0.83 (0.77–0.89)
2011	Murad et al., 2011; J. Clin. Endocrinol. Metab.	−14% with vitamin D supplementation	OR = 0.86 (0.77–0.96)
2014	Bolland et al., 2014; Lancet Diab. Endocrinol.	−5% with vitamin D supplementation	RR = 0.95 (0.90–1.00) *any quality trial*

OR, Odds ratio; RR, relative risk; RaR, rate ratio.

muscle weakness with vitamin D deficiency myopathy and Type II muscle atrophy has been shown to be reversible with treatment. Based on evidence from three smaller clinical trials among older adults (Ceglia et al., 2013; Sato et al., 2005; Sorensen et al., 1979), treatment with 1-alpha-calcidiol (Sorensen et al., 1979), vitamin D2 (Sato et al., 2005) or vitamin D3 (Ceglia et al., 2013) increased the number and diameter of Type II fast muscle fibres. Notably, in the most recent RCT comparing daily 4000 IU vitamin D3 with placebo over 4 months among 20 post-menopausal women, both VDR expression and Type II fibre diameter and number increased significantly in the vitamin D treated women and preferentially in the Type II fast muscle fibres (Ceglia et al., 2013).

4.6 Vitamin D and fall prevention

With evidence from several double-blind RCTs summarised in two trial-based and one pooled individual data meta-analyses, supplementation *with 800 IU vitamin D per day* should reduce the risk of falls (Bischoff-Ferrari et al., 2009a, 2011) (−19% to −34%) and hip (Bischoff-Ferrari et al., 2009b, 2012) (−30%) and any non-vertebral (Bischoff-Ferrari et al., 2009b, 2012) (−14%) fractures among seniors at risk of vitamin D deficiency. Notably, lower doses of vitamin D were not effective for both fall (Bischoff-Ferrari et al., 2009a) (<700 IU/day) and fracture (Bischoff-Ferrari et al., 2012) (<792 IU/day) prevention. Extending to more comprehensive meta-analyses of any vitamin D dose, route of application (oral daily, bolus, intra-muscular) and trial quality with respect to both study design (blinded and open designs) and fall ascertainment (any interval, prospective and retrospective), vitamin D supplementation still reduced fall risk significantly in most peer-reviewed meta-analyses (Table 4.1), including trials that tested active vitamin D metabolites.

The most recent 2014 comprehensive meta-analysis by Bolland and Reid included 25 clinical trials and documented only a small non-significant 5% benefit on fall prevention with vitamin D (Bolland et al., 2014). We suggest the main inconsistency in this recent meta-analysis (Bolland et al., 2014) compared with other meta-analyses presented in Table 4.1 is based on the inclusion of several trials that cannot be considered reliable indicators of true treatment efficacy. The authors included several open-design trials and trials that did not define how falls were assessed and/or did not assess falls prospectively or falls were imputed from fracture data. Also, the authors included trials where vitamin D was a component of a multi-factorial intervention, or was applied to seniors with unstable health such as those in acute hospital care or those who had suffered a stroke or heart failure. Further, they included several trials with oral or injected un-physiologically large bolus doses of 300,000 IU vitamin D or more, which may not be considered equivalent to daily or lower bolus dosing. In summary, the latest meta-analysis of Bolland and Reid does not invalidate the overall finding that vitamin D, at a daily dose of 800 IU per day, reduces the risk of falling among seniors at risk for vitamin D deficiency, both in the community and institutions (Bischoff-Ferrari et al., 2009a).

4.7 Fall reduction by type of vitamin D, by gender, and by dwelling

The 2009 meta-analysis of double-blind RCTs explored the benefit of the higher dose vitamin D (700–1000 IU per day) in several subgroups (Bischoff-Ferrari et al., 2009a): the pooled RR was 12% in trials that used vitamin D2 compared with 26% for trials that used vitamin D3, suggesting that vitamin D3 may be more effective in fall prevention. The effect of vitamin D in women suggested a pooled RR reduction of 15%, compared with 13% in the group as a whole, supporting a similar benefit among men. Finally, anti-fall benefits of 700–1000 IU vitamin D per day were present in subgroups of ambulatory and institutionalised senior adults as well as trials with a mean participant age 65–80 years or older. These findings were very consistent with the subgroup analyses explored for hip fracture risk reduction at a median dose of 800 IU in a pooled analysis of over 30,000 seniors enrolled in double-blind RCT of vitamin D and fracture reduction (Bischoff-Ferrari et al., 2012).

4.8 Fall reduction with 800 IU vitamin D by baseline 25-hydroxyvitamin D status

A 2-year placebo-controlled trial published by Uusi-Rasi et al. (2015) from Finland tested 800 IU vitamin D per day in vitamin D replete community-dwelling women age 70–80 years with a fall in the previous year. In the intent-to-treat analyses vitamin D maintained femoral neck bone mineral density and increased tibial trabecular density marginally, but did not reduce falls. The lack of benefit in the vitamin D group may be explained by the fact that none of the participants were vitamin D deficient at baseline. The placebo group started with mean 25(OH)D levels of 27.5 ng/mL and stayed at 27.5 ng/mL at the 2-year follow-up examination. The group that received vitamin D started at 25.1 ng/mL and reached 37.0 ng/mL at the 2-year follow-up. In contrast, 160 Brazilian post-menopausal women with a fall in the past year and low 25(OH)D levels, were treated with placebo or 1000 IU of vitamin D3 for 9 months (Cangussu et al., 2016). The vitamin D3 group started at a 25(OH)D level of 15.0 ng/mL and reached 27.5 ng/mL on treatment. The placebo group started at 16.9 ng/mL and dropped to 13.8 ng/mL during the trial. Over the 9-month intervention period, the risk of falling in the placebo group was almost twice that in the vitamin D treated group [RR 1.95 (1.23–3.08)] (Cangussu et al., 2016). Supporting a concept that fall prevention benefits of vitamin D supplementation are primarily seen in vitamin D deficient seniors, a recent meta-analysis of 17 RCTs suggested that a benefit of vitamin D on lower extremity strength may be seen primarily among those with vitamin D deficiency, defined as 25(OH)D < 10 ng/mL (Stockton et al., 2011).

4.9 Fall prevention and higher bolus doses of vitamin D

In the most recent higher-dose trial published in 2016, participants were randomised into three treatment groups: the first received a standard dose of 24,000 IU of vitamin D per month, the second received 60,000 IU of vitamin D per month and the third received 24,000 IU of vitamin D plus 300 µg of calcifediol per month. The 200 enrolled participants had fallen at least once in the 12 months leading up to the study, were on average 78 years old, 58% were vitamin D deficient [25(OH)D < 20 ng/mL], and all lived independently at home (Bischoff-Ferrari et al., 2016). Of the 200 participants, 60.5% (121 of 200) fell during the 12-month treatment period. The 2-monthly high-dose groups did not improve lower extremity function superior to the standard-monthly dose of 24,000 IU vitamin D and had higher percentages of participants who fell (66.9% and 66.1%, respectively) compared with the 24,000 IU group (47.9%). Participants in the 24,000 IU vitamin D group (equivalent to 800 IU/day) experienced the most improved lower extremity function and also had the fewest number of falls. A consistent pattern was seen by achieved 25-hydroxyvitamin D blood levels. The best functional improvement and fewest falls were observed at the lower replete 25-hydroxyvitamin D quartile range of 21.3–30.3 ng/mL, while no functional benefit plus most falls were observed at the highest achieved 25-hydroxyvitamin D quartile range (44.7–98.9 ng/mL). Notably, for the dosages examined in this study, the 2 higher doses were most likely to achieve the detrimental highest quartile at 6 months and 12 months' follow-up, independent of starting level. On the other hand, participants in the standard dose of 24,000 IU vitamin D were most likely to achieve the optimal lower replete range of 21.3–30.3 ng/mL, and none of the participants of this group reached the undesirable highest quartile range of greater than 45 ng/mL.

High oral doses of vitamin D and fall prevention have been evaluated in two other trials with prospective fall assessment (Bischoff-Ferrari et al., 2010; Sanders et al., 2010) and one trial that assessed falls retrospectively (Trivedi et al., 2003). In one trial among 173 frail seniors after acute hip fracture, 2000 IU vitamin D/day vs. 800 IU/day did not improve lower extremity function or reduce falls over a 12 month follow-up (+28%; 95% CI: −4%, +68%) (Bischoff-Ferrari et al., 2010). Notably, the achieved mean 25(OH)D level at 12 months was 44.6 ng/mL in the 2000 IU/day group compared with 35.4 ng/mL in the 800 IU/day group (Bischoff-Ferrari et al., 2010). In another trial with 2256 senior women at high risk for hip fracture, an annual bolus of 500,000 IU vitamin D vs. placebo increased risk of falling (RR = 1.15; 95% CI: 1.02–1.30) (Sanders et al., 2010). The bolus group achieved a 25(OH)D level of 48 ng/mL at 1 month and 36 ng/mL at 3 months follow-up, the timeframe where most falls occurred in the trial (Sanders et al., 2010). Notably, in both of these trials, high dose vitamin D shifted participants to a mean 25(OH)D level overlapping the highest achieved quartile in the 2016 trial outlined previously (Bischoff-Ferrari et al., 2010; Sanders et al., 2010). In a third trial of 2686 community-dwelling seniors, age 65–85 years, a bolus of 100,000 IU every 4 months for 5 years, reduced the risk of any new fracture significantly by 22%, but did not reduce the risk of falling (RR = 0.93; 95% CI: 0.76–1.14) (Trivedi et al., 2003). However, falls were only assessed retrospectively for the last year of follow-up and the treatment group shifted to a mean 25(OH)D level of 30 ng/mL (Trivedi et al., 2003).

Notably, the physiology behind a possible detrimental effect of a high bolus dose of vitamin D on muscle functions and falls remains unclear and needs further investigation. Two ongoing trials (VITAL and DO-HEALTH) use 2000 IU vitamin D per day. These trials will provide important opportunities to verify and expand higher dose findings to other endpoints.

4.10 Current guidelines on vitamin D and fall prevention

While the US Institutes of Medicine (IOM) recommendation of an increase in vitamin D intake is supported by the available data from double-blind RCTs of fracture risk, a threshold of 20 ng/mL for 25(OH)D blood concentration is not. In two 2009 meta-analyses of double-blind RCTs, a threshold of 20 ng/mL was insufficient for fracture or fall reduction based on achieved 25(OH)D levels in the treatment groups (Bischoff-Ferrari et al., 2009a,b) with optimal levels suggested to be 24 ng/mL for fall prevention and 30 ng/mL for fracture prevention. Accordingly, and in contrast to the IOM report, the International Osteoporosis Foundation (IOF) recommended in their 2010 position paper on vitamin D a threshold of 30 ng/mL for optimal fall and fracture reduction and recommended 800–1000 IU vitamin D per day for seniors age 60 years and older (Dawson-Hughes et al., 2010).

Regarding falls, the IOM synopsis was that the evidence of vitamin D on fall prevention is inconsistent. This is in contrast to the 2011 assessment of the Agency for Healthcare Research and Quality (AHRQ) for the US Preventive Services Task Force (Michael et al., 2011), the 2010 American Geriatric Society/British Geriatric Society Clinical Practice Guideline (AGS/BGS, 2011), and to the 2010 assessment by the IOF (Dawson-Hughes et al., 2010), all three of which identified vitamin D as an effective intervention to prevent falling in older adults.

Also, the most recent EFSA recommendations on dietary reference values for vitamin D (EFSA, 2016) concluded that current evidence is suggestive of a beneficial effect of vitamin D on falls. However, their recommendation of an adequate intake was set at 600 IU (15 μg) per day for all adults with no higher requirement for older adults. Notably, less than 800 IU per day for older adults has been shown to be insufficient for fall (Bischoff-Ferrari et al., 2009a, 2012) and fracture prevention (Bischoff-Ferrari et al., 2012) among older adults at risk of vitamin D deficiency. The Scientific Advisory Committee on Nutrition in the United Kingdom set the lowest vitamin D dietary reference value for all adults with no additional requirement for older adults at 400 IU per day (SACN, 2016). All other international recommendations support 800 IU vitamin D per day for adults age 60 or 65 years and older.

4.11 Authorised EU health claim

In 2014, a reduction of disease risk health claim was authorised for use in the European Union relating to the role of vitamin D in reducing the risk of falls. As outlined at the beginning of this chapter, falls are considered a key risk factor for fractures. The target group for the claim is men and women aged 60 years and over.

EFSA had previously published its opinion on the claim application in 2011 (EFSA, 2011). In this opinion, EFSA concluded that falling is a risk factor for bone fracture, and that there was sufficient evidence of cause and effect to justify the claim. Since EFSA's opinion on the level of scientific evidence was published in 2011, we reviewed evidence relevant to this claim after its publication.

EFSA proposed in 2011, a daily intake of 800 IU vitamin D for fall prevention. This was consistent with an EFSA opinion on the daily intake of vitamin D to reduce loss of bone mineral density and thus reduce the risk of osteoporotic fractures in postmenopausal women aged 50 years and over, which was set at 800 IU vitamin D per day from all sources (EFSA, 2010). As the daily intake required to produce the beneficial effect was higher than usual dietary recommendations, the EU Commission asked EFSA to review the Tolerable Upper Intake Level (UL) for vitamin D. This opinion was issued in 2012 (EFSA, 2012) and EFSA revised the UL from the previous value of 2000–4000 IU for adults, consistent with the 2010 report by the IOM.

Eventually the claim was authorised with the wording and conditions of use as described in Table 4.2. Notably, due to the high daily intake of vitamin D required for the beneficial effect on falls and fracture, the claim is only authorised for use on food supplements. The conditions of use also allow for the additional daily intake of vitamin D from natural food sources. Table 4.2 also details the other authorised health claims for vitamin D.

4.12 Conclusions

Overall, there is consistent mechanistic and clinical evidence, that vitamin D is important in muscle health and reduces fall risk among seniors at risk for vitamin D deficiency. Most evidence has been collected for daily vitamin D at a dose of 800 IU per day, which is the recommended dose in most international guidelines on vitamin D targeting all subgroups in the senior population. Regarding higher bolus doses of vitamin D, 24,000 IU monthly (equivalent to 800 IU daily) has been safe and effective in one recent trial (Bischoff-Ferrari et al., 2016), while higher monthly doses of vitamin D (60,000 IU) or a combination with calcifediol may not be warranted in seniors with a prior fall due to a potentially deleterious effect on falls (Bischoff-Ferrari et al., 2016). Also high-annual doses (500,000 IU) cannot be recommended as a safe strategy of fall prevention (Sanders et al., 2010). For a target 25(OH)D level for optimal fall prevention, recent data support that there may be a therapeutic range for vitamin D with both too low (<20 ng/mL) and too high (>45 ng/mL) 25(OH)D levels of vitamin D (Bischoff-Ferrari et al., 2016). Notably, both daily 800 IU and monthly

Table 4.2 Authorised EU health claims for vitamin D

Type of EU health claim	Authorised claim wording	Conditions of use
Article 14.1(a) disease risk reduction claim	Vitamin D helps to reduce the risk of falling associated with postural instability and muscle weakness. Falling is a risk factor for bone fractures among men and women 60 years of age and older	The claim may be used only for food supplements, which provide at least 15 μg of vitamin D per daily portion. Information shall be given to the consumer that the beneficial effect is obtained with a daily intake of 20 μg of vitamin D from all sources. For food supplements with added vitamin D the claim may be used only for those targeting men and women 60 years and older
Article 13.1 general function claim	Vitamin D contributes to normal absorption/utilisation of calcium and phosphorus	The claim may be used only for food which is at least a source of vitamin D as referred to in the claim SOURCE OF [NAME OF VITAMIN/S] AND/OR [NAME OF MINERAL/S] as listed in the Annex to Regulation (EC) No 1924/2006
Article 13.1 general function claim	Vitamin D contributes to normal blood calcium levels	The claim may be used only for food which is at least a source of vitamin D as referred to in the claim SOURCE OF [NAME OF VITAMIN/S] AND/OR [NAME OF MINERAL/S] as listed in the Annex to Regulation (EC) No 1924/2006
Article 13.1 general function claim	Vitamin D contributes to the maintenance of normal bones	The claim may be used only for food which is at least a source of vitamin D as referred to in the claim SOURCE OF [NAME OF VITAMIN/S] AND/OR [NAME OF MINERAL/S] as listed in the Annex to Regulation (EC) No 1924/2006
Article 13.1 general function claim	Vitamin D contributes to the maintenance of normal muscle function	The claim may be used only for food which is at least a source of vitamin D as referred to in the claim SOURCE OF [NAME OF VITAMIN/S] AND/OR [NAME OF MINERAL/S] as listed in the Annex to Regulation (EC) No 1924/2006

Article 13.1 general function claim	Vitamin D contributes to the maintenance of normal teeth	The claim may be used only for food which is at least a source of vitamin D as referred to in the claim SOURCE OF [NAME OF VITAMIN/S] AND/OR [NAME OF MINERAL/S] as listed in the Annex to Regulation (EC) No 1924/2006
Article 13.1 general function claim	Vitamin D contributes to the normal function of the immune system	The claim may be used only for food which is at least a source of vitamin D as referred to in the claim SOURCE OF [NAME OF VITAMIN/S] AND/OR [NAME OF MINERAL/S] as listed in the Annex to Regulation (EC) No 1924/2006
Article 13.1 general function claim	Vitamin D has a role in the process of cell division	The claim may be used only for food which is at least a source of vitamin D as referred to in the claim SOURCE OF [NAME OF VITAMIN/S] AND/OR [NAME OF MINERAL/S] as listed in the Annex to Regulation (EC) No 1924/2006
Article 14.1(b) children's claim	Vitamin D is needed for normal growth and development of bone in children	The claim can be used only for food which is at least a source of Vitamin D as referred to in the claim SOURCE OF [NAME OF VITAMIN/S] AND/OR [NAME OF MINERAL/S] as listed in the Annex to Regulation 1924/2006
Article 14.1(b) children's claim	Vitamin D contributes to the normal function of the immune system in children	The claim may be used only for food which is at least a source of vitamin D as referred to in the claim SOURCE OF [NAME OF VITAMIN/S] AND/OR [NAME OF MINERAL/S] as listed in the Annex to Regulation (EC) No 1924/2006

24,000 IU provide a very high-safety profile to shift 25(OH)D levels to a desirable range between 20 and 30 ng/mL, independent of baseline 25(OH)D level (Bischoff-Ferrari et al., 2016) without risk of shifting senior adults to levels above 45 ng/mL (Bischoff-Ferrari et al., 2016).

Relevant units:

$$\text{Vitamin D}: \quad 400\ \text{IU} = 10\ \mu\text{g};\ 800\ \text{IU} = 20\ \mu\text{g}$$

$$25(\text{OH})\text{D}: \quad 20\ \text{ng/mL} = 50\ \text{nmol/L};\ 30\ \text{ng/mL} = 75\ \text{nmol/L}$$

4.13 Sources of further information

EU Register of Nutriiton and Health Claims Made on Foods: (http://ec.europa.eu/food/safety/labelling_nutrition/claims/register/public/?event=register.home).

IoM (Institute of Medicine), 2010. Dietary reference intakes for calcium and vitamin D. In: Ross, A.C., Taylor, C.L., Yaktine, A.L., Del Valle, H.B. (Eds.), The National Academies Press, Washington, DC.

References

AGS/BGS, 2011. American Geriatric Society/British Geriatric Society Guidelines on Fall Prevention in Older Persons. Available from: http://www.patientsafetysolutions.com/docs/February_2011_Updated_Fall_Prevention_Guidelines.htm.

Al-Shoha, A., Qiu, S., Palnitkar, S., Rao, D.S., 2009. Osteomalacia with bone marrow fibrosis due to severe vitamin D deficiency after a gastrointestinal bypass operation for severe obesity. Endocr. Pract. 15 (6), 528–533.

Bischoff, H.A., Stahelin, H.B., Dick, W., et al., 2003. Effects of vitamin D and calcium supplementation on falls: a randomized controlled trial. J. Bone Miner. Res. 18, 343–351.

Bischoff-Ferrari, H.A., 2012. Relevance of vitamin D in muscle health. Rev. Endocr. Metab. Disord. 13, 71–77.

Bischoff-Ferrari, H.A., Borchers, M., Gudat, F., Durmuller, U., Stahelin, H.B., Dick, W., 2004a. Vitamin D receptor expression in human muscle tissue decreases with age. J. Bone Miner. Res. 19, 265–269.

Bischoff-Ferrari, H.A., Dawson-Hughes, B., Orav, E.J., et al., 2016. Monthly high-dose vitamin D treatment for the prevention of functional decline: a randomized clinical trial. JAMA Intern. Med. 176, 175–183.

Bischoff-Ferrari, H.A., Dawson-Hughes, B., Platz, A., et al., 2010. Effect of high-dosage cholecalciferol and extended physiotherapy on complications after hip fracture: a randomized controlled trial. Arch. Intern. Med. 170, 813–820.

Bischoff-Ferrari, H.A., Dawson-Hughes, B., Staehelin, H.B., et al., 2009a. Fall prevention with supplemental and active forms of vitamin D: a meta-analysis of randomised controlled trials. BMJ 339, b3692.

Bischoff-Ferrari, H.A., Dawson-Hughes, B., Willett, W.C., et al., 2004b. Effect of vitamin D on falls: a meta-analysis. JAMA 291 (16), 1999–2006.

Bischoff-Ferrari, H.A., Dietrich, T., Orav, E.J., Dawson-Hughes, B., 2004c. Positive association between 25-hydroxy vitamin D levels and bone mineral density: a population-based study of younger and older adults. Am. J. Med. 116, 634–639.

Bischoff-Ferrari, H.A., Dietrich, T., Orav, E.J., et al., 2004d. Higher 25-hydroxyvitamin D concentrations are associated with better lower-extremity function in both active and inactive persons aged ≥60 years. Am. J. Clin. Nutr. 80, 752–758.

Bischoff-Ferrari, H.A., Orav, E.J., Dawson-Hughes, B., 2006. Effect of cholecalciferol plus calcium on falling in ambulatory older men and women: a 3-year randomized controlled trial. Arch. Intern. Med. 166, 424–430.

Bischoff-Ferrari, H.A., Willett, W.C., Orav, E.J., Kiel, D.P., Dawson-Hughes, B., 2011. Re: Fall prevention with Vitamin D. Clarifications needed. Available from: http://www.bmj.com/content/339/bmj.b3692/rapid-responses.

Bischoff-Ferrari, H.A., Willett, W.C., Orav, E.J., et al., 2012. A pooled analysis of vitamin D dose requirements for fracture prevention. N. Engl. J. Med. 367, 40–49.

Bischoff-Ferrari, H.A., Willett, W.C., Wong, J.B., et al., 2009b. Prevention of nonvertebral fractures with oral vitamin D and dose dependency: a meta-analysis of randomized controlled trials. Arch. Intern. Med. 169, 551–561.

Bolland, M.J., Grey, A., Gamble, G.D., Reid, I.R., 2014. Vitamin D supplementation and falls: a trial sequential meta-analysis. Lancet Diab. Endocrinol. 2, 573–580.

Bouillon, R., Bischoff-Ferrari, H., Willett, W., 2008. Vitamin D and health: perspectives from mice and man. J. Bone Miner. Res. 23 (7), 974–979.

Broe, K.E., Chen, T.C., Weinberg, J., Bischoff-Ferrari, H.A., Holick, M.F., Kiel, D.P., 2007. A higher dose of vitamin d reduces the risk of falls in nursing home residents: a randomized, multiple-dose study. J. Am. Geriatr. Soc. 55, 234–239.

Cameron, I.D., Murray, G.R., Gillespie, L.D., et al., 2010. Interventions for preventing falls in older people in nursing care facilities and hospitals. Cochrane Database Syst. Rev. 20, CD005465.

Cangussu, L.M., Nahas-Neto, J., Orsatti, C.L., et al., 2016. Effect of isolated vitamin D supplementation on the rate of falls and postural balance in postmenopausal women fallers: a randomized, double-blind, placebo-controlled trial. Menopause 23 (3), 267–274.

CDC, Centers for Disease Control and Prevention, 2006. Fatalities and injuries from falls among older adults—United States, 1993–2003 and 2001–2005. MMWR Morbi. Mortal. Wkly. Rep. 55 (45), 1221–1224.

Ceglia, L., Niramitmahapanya, S., da Silva Morais, M., et al., 2013. A randomized study on the effect of vitamin D3 supplementation on skeletal muscle morphology and vitamin D receptor concentration in older women. J. Clin. Endocrinol. Metab. 98, E1927–E1935.

Cummings, S.R., Nevitt, M.C., Browner, W.S., et al., 1995. Risk factors for hip fracture in white women. Study of Osteoporotic Fractures Research Group. N. Engl. J. Med. 332, 767–773.

Dawson-Hughes, B., Harris, S.S., Krall, E.A., Dallal, G.E., 1997. Effect of calcium and vitamin D supplementation on bone density in men and women 65 years of age or older. N. Engl. J. Med. 337, 670–676.

Dawson-Hughes, B., Mithal, A., Bonjour, J.P., et al., 2010. IOF position statement: vitamin D recommendations for older adults. Osteoporos. Int. 21, 1151–1154.

Eberstadt, N., Groth, H., 2007. Europe's coming demographic challenge: unlocking the value of health. American Enterprise Institute for Health Policy Research, Washington, D.C.

EFSA, 2009. Scientific Opinion on the substantiation of health claims related to calcium and vitamin D and maintenance of bone (ID 350) pursuant to Article 13(1) of Regulation (EC) No 1924/2006.

EFSA, 2010. Scientific opinion in relation to the authorisation procedure for health claims on calcium and vitamin D and the reduction of the risk of osteoporotic fractures by reducing bone loss pursuant to Article 14 of Regulation (EC) No 1924/2006. EFSA J. 8 (5), 1609.

EFSA, 2011. Scientific opinion on the substantiation of a health claim related to vitamin D and risk of falling pursuant to Article 14 of Regulation (EC) No 1924/2006. Scientific Opinion of the Panel on Dietetic Products, Nutrition and Allergies. EFSA J. 9 (9), 2382.

EFSA, 2012. Scientific opinion on the tolerable upper intake level of vitamin D. EFSA J. 10 (7), 2813.

EFSA, 2016. Dietary reference values of vitamin D. EFSA J. 14 (10), 4547.

Endo, I., Inoue, D., Mitsui, T., et al., 2003. Deletion of vitamin D receptor gene in mice results in abnormal skeletal muscle development with deregulated expression of myoregulatory transcription factors. Endocrinology 144, 5138–5144.

Englander, F., Hodson, T.J., Terregrossa, R.A., 1996. Economic dimensions of slip and fall injuries. J. Forensic Sci. 41, 733–746.

Grimby, G., Saltin, B., 1983. The ageing muscle. Clin. Physiol. 3, 209–218.

Gullberg, B., Johnell, O., Kanis, J.A., 1997. World-wide projections for hip fracture. Osteoporos. Int. 7, 407–413.

Jackson, C., Gaugris, S., Sen, S.S., Hosking, D., 2007. The effect of cholecalciferol (vitamin D3) on the risk of fall and fracture: a meta-analysis. QJM 100, 185–192.

Kalyani, R.R., Stein, B., Valiyil, R., Manno, R., Maynard, J.W., Crews, D.C., 2010. Vitamin D treatment for the prevention of falls in older adults: systematic review and meta-analysis. J. Am. Geriatr. Soc. 58, 1299–1310.

Melton, 3rd., L.J., Crowson, C.S., O'Fallon, W.M., 1999. Fracture incidence in Olmsted County, Minnesota: comparison of urban with rural rates and changes in urban rates over time. Osteoporos. Int. 9, 29–37.

Michael, Y.L., Whitlock, E.P., Lin, J.S., Fu, R., O'Connor, E.A., Gold, R., 2011. Primary care-relevant interventions to prevent falling in older adults: a systematic evidence review for the US Preventive services task force. Ann. Intern. Med. 153, 815–825.

Muir, S.W., Montero-Odasso, M., 2011. Effect of vitamin D supplementation on muscle strength, gait and balance in older adults: a systematic review and meta-analysis. J. Am. Geriatr. Soc. 59, 2291–2300.

Murad, M.H., Elamin, K.B., Abu Elnour, N.O., et al., 2011. Clinical review: the effect of vitamin D on falls: a systematic review and meta-analysis. J. Clin. Endocrinol. Metab. 96, 2997–3006.

O'Donnell, S., Moher, D., Thomas, K., Hanley, D.A., Cranney, A., 2008. Systematic review of the benefits and harms of calcitriol and alfacalcidol for fractures and falls. J. Bone Miner. Metab. 26, 531–542.

Pfeifer, M., Begerow, B., Minne, H.W., Abrams, C., Nachtigall, D., Hansen, C., 2000. Effects of a short-term vitamin D and calcium supplementation on body sway and secondary hyperparathyroidism in elderly women. J. Bone Miner. Res. 15, 1113–1118.

Pfeifer, M., Begerow, B., Minne, H.W., Suppan, K., Fahrleitner-Pammer, A., Dobnig, H., 2009. Effects of a long-term vitamin D and calcium supplementation on falls and parameters of muscle function in community-dwelling older individuals. Osteoporos. Int. 20 (2), 315–322.

Pluijm, S.M., Smit, J.H., Tromp, E.A., et al., 2006. A risk profile for identifying community-dwelling elderly with a high risk of recurrent falling: results of a 3-year prospective study. Osteoporos. Int. 17, 417–425.

Richy, F., Dukas, L., Schacht, E., 2008. Differential effects of D-hormone analogs and native vitamin D on the risk of falls: a comparative meta-analysis. Calcif. Tissue Int. 82, 102–107.

SACN, Scientific Advisory Committee on Nutrition (UK), 2016. Vitamin D and Health Report. Available from: http://ec.europa.eu/food/safety/labelling_nutrition/claims/register/public/?event=register.home.

Sanders, K.M., Stuart, A.L., Williamson, E.J., et al., 2010. Annual high-dose oral vitamin D and falls and fractures in older women: a randomized controlled trial. JAMA 303, 1815–1822.

Sato, Y., Iwamoto, J., Kanoko, T., Satoh, K., 2005. Low-dose vitamin D prevents muscular atrophy and reduces falls and hip fractures in women after stroke: a randomized controlled trial. Cerebrovasc. Dis. 20, 187–192.

Schott, G.D., Wills, M.R., 1976. Muscle weakness in osteomalacia. Lancet 1, 626–629.

Sorensen, O.H., Lund, B., Saltin, B., et al., 1979. Myopathy in bone loss of ageing: improvement by treatment with 1 alpha-hydroxycholecalciferol and calcium. Clin. Sci. (Lond.) 56, 157–161.

Srikuea, R., Zhang, X., Park-Sarge, O.K., Esser, K.A., 2012. VDR and CYP27B1 are expressed in C2C12 cells and regenerating skeletal muscle: potential role in suppression of myoblast proliferation. Am. J. Physiol. Cell Physiol. 303, C396–C405.

Stevens, J.A., Corso, P.S., Finkelstein, E.A., Miller, T.R., 2006. The costs of fatal and nonfatal falls among older adults. Inj. Prev. 12, 290–295.

Stockton, K.A., Mengersen, K., Paratz, J.D., Kandiah, D., Bennell, K.L., 2011. Effect of vitamin D supplementation on muscle strength: a systematic review and meta-analysis. Osteoporos. Int. 22, 859–871.

Tinetti, M.E., 1994. Prevention of falls and fall injuries in elderly persons: a research agenda. Prev. Med. 23, 756–762.

Tinetti, M.E., Williams, C.S., 1997. Falls, injuries due to falls, and the risk of admission to a nursing home. N. Engl. J. Med. 337, 1279–1284.

Trivedi, D.P., Doll, R., Khaw, K.T., 2003. Effect of four monthly oral vitamin D3 (cholecalciferol) supplementation on fractures and mortality in men and women living in the community: randomised double blind controlled trial. BMJ 326, 469.

Uusi-Rasi, K., Patil, R., Karinkanta, S., et al., 2015. Exercise and vitamin D in fall prevention among older women: a randomized clinical trial. JAMA Intern. Med. 175 (5), 703–711.

Visser, M., Deeg, D.J., Puts, M.T., Seidell, J.C., Lips, P., 2006. Low serum concentrations of 25-hydroxyvitamin D in older persons and the risk of nursing home admission. Am. J. Clin. Nutr. 84, 616–622.

Wicherts, I.S., van Schoor, N.M., Boeke, A.J., et al., 2007. Vitamin D status predicts physical performance and its decline in older persons. J. Clin. Endocrinol. Metab. 92 (6), 2058–2065.

Yoshikawa, S., Nakamura, T., Tanabe, H., Imamura, T., 1979. Osteomalacic myopathy. Endocrinol. Jpn. 26, 65–72.

Authorised EU health claim for supplementary folic acid

5

Michele J. Sadler
Rank Nutrition Ltd, Bethersden, Kent, United Kingdom

5.1 Introduction

The relationship between adequate intake of folate, including folic acid, prior to and during the early weeks of pregnancy to reduce the risk of neural tube defects (NTDs) such as spina bifida in the foetus is long established. Taking a daily 400-μg folic acid supplement during this vulnerable period is a global WHO recommendation (WHO, 2006, 2012). As the number of NTD-affected pregnancies remains stubbornly high in Europe, resulting in severe disability to infants born with the condition and unbearable distress to mothers who may choose to terminate their pregnancy, there is an ongoing need to ensure that women of childbearing age are aware of and act upon this recommendation.

As NTDs occur within the first 28 days of pregnancy, which is often before a woman knows she is pregnant, the timing of folic acid supplementation is crucial. As a high proportion of pregnancies are unplanned, the advice should ideally be targeted at women who might become pregnant or are of childbearing age and not just those planning a pregnancy. Advice is also for taking folic acid supplementation prior to and during the first 12 weeks of pregnancy. Advice on the point at which folic acid supplementation should be started when planning a pregnancy varies across Europe (Cawley et al., 2016). However, the take-up of supplementation is low; even if women are aware of the advice, its relevance is not always apparent to those not planning pregnancy and, hence, motivation to comply is low.

A report published in 2011 'Act against Europe's most common birth defects: one year on—Defining Neural Tube Defect prevention strategies in Europe' identified that there are 4500 NTD pregnancies each year in Europe (Joint Report on NTD, 2011). Though the condition occurs in the initial stages of pregnancy, it is not usually diagnosed until the 20-week scan. An estimated 72% of affected pregnancies are terminated following prenatal diagnosis, causing much distress to the mothers and their families.

Scientific evidence suggests that up to 70% of NTDs could be avoided by the maintenance of adequate folate status in women of childbearing age prior to conception. A European Commission publication 'Communication of Rare Diseases: Europe's Challenges' (EC, 2008) recognised that NTDs are one of the few rare diseases that could, in many cases, be prevented and that the protective measure is very straightforward, the intake of folic acid supplements at the appropriate time. Hence all available means to communicate this crucial message should be utilised.

Foods, Nutrients and Food Ingredients with Authorised EU Health Claims. http://dx.doi.org/10.1016/B978-0-08-100922-2.00005-X
Copyright © 2018 Elsevier Ltd. All rights reserved.

It was recognised that a European-wide Article 14.1(a), Reduction of Disease Risk Claim, under the Nutrition and Health Claims Regulation (EU, 2006b), would provide an additional means of raising awareness of the recommendation among women of childbearing age with the aim of increasing uptake of folic acid supplementation. Prior to authorisation of the claim, it was not possible to completely indicate why folic acid supplementation is needed prior to and in the early stages of pregnancy, as any reference to NTDs on product labels and in advertising would be classed as a medicinal claim. Hence a reduction of disease risk claim, based on the risk factor for the condition, would allow communication of a stronger health message.

To achieve this important objective, an application was made for the European Union to enable wider communication of this recommendation. A dossier was written and the application was submitted by Rank Nutrition Ltd. It was funded by three dietary supplement trade associations in the United Kingdom [Health Food Manufacturer's Association (HFMA), Proprietary Association of Great Britain (PAGB) and Council for Responsible Nutrition, United Kingdom (CRN-UK)]. These associations came together in 2012 in conjunction with the United Kingdom's spina bifida and hydrocephalus charity (Shine). They all shared a common concern that further to publication of the authorised list of health claims in Europe (EU, 2012), a glaring omission was a claim relating to the benefit of maternal intake of folic acid and reduced risk of NTDs despite conclusive evidence for this health benefit. The wording of a related authorised Article 13.1 health claim for folate allowed only the statement that 'Folate contributes to normal maternal tissue growth during pregnancy'. Even though the European Food Safety Authority (EFSA) opinion for this claim referred to the need of folate for the growth of the foetus in addition to maternal tissues such as the placenta, uterus and blood cell mass, and to the role of folate in cell division (EFSA, 2009b), reference to the foetus was not possible in the context of an Article 13.1 health claim.

After taking early advice from the UK Department of Health (DH), a regulatory dossier was compiled and submitted via the UK competent authority.

5.2 What are neural tube defects?

NTDs are congenital malformations that occur during the structural development of the neural tube and are caused by failure of the neural tube to close properly. The development and closure of the neural tube, which forms into the brain and spinal cord, is normally completed within 21–28 days after conception. Causes of NTDs are multifactorial and can be due to genetic factors, environmental risk factors or maternal conditions such as low status of folate and vitamin B12.

Identification and classification of NTDs in live births, terminations, foetal death or miscarriage are determined according to the International Classification of Diseases (ICD) codings which is a valid methodology. NTDs include spina bifida (ICD 9 code 741), resulting in a cleft spine; anencephaly (ICD 9 code 740), arising when the cephalic end of the neural tube fails to close; and encephalocoele (ICD 9 code 742), arising when there is an opening in one or more vertebrae. NTDs can lead to miscarriage, neonatal death or, should the infant survive, lifelong disability. In view of the disability associated with NTDs, identification of affected pregnancies will often result in the decision to terminate the pregnancy.

The critical period to ensure adequacy of maternal folate status is before closure of the neural tube, which occurs in the first 28 days following conception. However, at such an early stage of pregnancy, a woman is unlikely to know that she is pregnant. Hence preconceptional supplementation with folic acid is crucial to reduce the risk.

5.3 Characterisation of folic acid

The health claim is authorised for folic acid [pteroylglutamic acid (PGA)] which is a member of the B family of vitamins. Folate is the generic name for the vitamin, and folic acid is a synthetic form that is used in food supplements and in food fortification, that is not a natural component of the diet. In contrast to folate, folic acid is an oxidised compound with one glutamate residue. It becomes biologically active after in vivo reduction and is highly stable and more bioavailable relative to natural folates. Natural dietary folates are mostly derivatives of tetrahydrofolate (THF), which are reduced folates. Folate and its related compounds consist of an aromatic pteridine ring linked to *p*-aminobenzoic acid and one or more (usually 5–7) glutamate residues. As such they are unstable, and losses occur in the presence of oxygen due to light and at high temperatures such as those encountered in cooking and food processing (EFSA, 2014).

The health claim applies to all forms of folate authorised for addition to foods and food supplements, that is to folic acid. The authorised health claim is unusual in the sense that it is one of three reduction of disease risk claims authorised only for food supplements. This reflects that obtaining the effective intake of some nutrients, particularly those present in relatively few foods, is best achieved by the intake of food supplements. This ensures that the regulatory requirement in Article 5.1(d) of Regulation EU 1924/2006 is met (EU, 2006b), whereby 'the quantity of the product that can reasonably be expected to be consumed provides a significant quantity of the nutrient or other substance to which the claim relates'. The other two such claims are for vitamin D and reduced risk of falling, and for calcium and vitamin D and reduced loss of bone mineral in post-menopausal women (EFSA, 2009a, 2011), as discussed in the previous two chapters.

5.3.1 Bioavailability

Depending on the methodology used, estimates of the bioavailability of food folates relative to folic acid vary widely, and range from 10% to 98% lower (McNulty and Pentieva, 2004). A more recent 4-week intervention study concluded that the aggregate bioavailability of folates from fruit, vegetables and liver is approximately 80% that of folic acid (Winkels et al., 2007).

To attempt to adjust for differences in bioavailability between folic acid and naturally occurring folates, the term dietary folate equivalents (DFEs) was introduced in 1998. Based on data available at that time, DFE assumed 50% bioavailability of food folate, 85% bioavailability of folic acid when taken with food (either as a food supplement or from fortified food), and 100% for folic acid when taken on an empty stomach with water. On this basis the bioavailability of naturally occurring food folate is 60% of folic acid when consumed with a meal (Caudill, 2010).

Due to the serious nature of NTDs, it is important for women to feel reassured that sufficient folic acid has been both ingested and absorbed, particularly as there is a limited time period within which to ensure that the risk factor is favourable. Ingestion of a known amount of highly bioavailable folic acid in the form of a dietary supplement is the most reliable way for women to feel confident that they have consumed sufficient additional folic acid, and hence advice on supplementation is a key strategy for reducing risk of NTDs.

5.4 Formulating the health claim

The Nutrition and Health Claims Regulation (EU, 2006b) defines a reduction of disease risk claim as 'any health claim that states, suggests or implies that the consumption of a food category, a food or one of its constituents, significantly reduces a risk factor in the development of a human disease'. All of the reduction of disease risk claims that have been authorised to date follow a common pattern of wording:

[Subject of claim] 'has been shown to'/'help to' 'reduce/increase/neutralise' [risk factor]. [Risk factor] is a risk factor for [disease].

In addition to the large body of direct evidence that folic acid reduces the risk of NTDs, it was also necessary to identify the risk factor that folic acid supplementation beneficially alters, resulting in the reduction of NTD risk. The claim was thus based on folate status as the risk factor and NTDs as the disease.

5.5 Substantiating the health claim

In line with EFSA guidance, a systematic approach was taken to identify the scientific literature relevant to the claim. This included evidence from intervention trials, observational studies, systematic reviews and consensus reports. Most of papers identified were observational studies (cohort studies, case–control studies and cross-sectional studies), which provided valuable supporting evidence for the claim, while intervention studies provided the key substantiating evidence. In view of the large body of human data, animal studies were not reviewed.

5.5.1 Direct evidence for folic acid and reduced risk of NTDs

The strong body of direct evidence from primary intervention trials, taking supplementary folic acid prior to conception can help prevent both first occurrence and recurrence of NTDs (Table 5.1), that substantiates the beneficial effect of folic acid in reducing risk of NTDs in the foetus. This evidence also supports maternal folic acid status as an independent risk factor for NTDs.

An early double-blind randomised controlled trial (RCT) in women with a previous NTD-affected pregnancy had shown that, compared with placebo, a daily 4-mg folic acid supplement prior to conception and during the first 12 weeks of pregnancy significantly reduced the risk of recurrence of NTDs ($P = 0.04$) (Laurence et al., 1981). In this study there were no recurrences among the mothers who complied with the intervention (44 out of 60 women), two recurrences were among the non-compliant women and four recurrences were among the placebo group. An early multicentre, non-RCT had also demonstrated that taking 360 μg folic acid/day as part of a multivitamin supplement, for not less than 28 days before conception and during the first ~6 weeks of pregnancy, also reduced recurrence of NTDs (Smithells et al., 1983).

The definitive evidence that risk of NTDs is reduced by folic acid supplementation was provided in 1991 by a UK Medical Research Council (MRC) RCT that investigated risk of recurrence of NTDs (MRC, 1991), and by a further RCT in Hungary that confirmed a protective effect against a first occurrence of NTDs (Czeizel et al., 1994). The design of the MRC trial included four groups: 4 mg folic acid plus iron and calcium; 4 mg folic acid plus iron, calcium and other micronutrients (vitamins A, B2, B6, C, D and nicotinamide); and iron, calcium and other micronutrients (as before) without folic acid. The MRC trial design also included a control group taking iron and calcium without folic acid or other micronutrients. The supplements were taken before and after conception for up to 12 weeks of pregnancy in women who had previously had an NTD-affected pregnancy. Recurrence of NTDs was 72% lower in subjects receiving the folic acid intervention, establishing a protective effect of folic acid. The trial was stopped early, as there were sufficient conclusive results demonstrating a protective effect of folic acid, and it was deemed unethical for the control group to continue on the placebo.

In the study by Czeizel et al. (1994), women without a history of NTDs were supplemented daily with 800 μg folic acid and multiple micronutrients, compared with a control group of multiple micronutrients without folic acid, taken prior to conception and for the first 12 weeks of pregnancy. Though the evidence for reduction of NTDs was not statistically significant, there were no reported cases of NTDs in the folic acid group.

Later trials have added to the body of evidence for a protective effect of folic acid. Following publication of the MRC trial, the ethical committee of a similar trial in India also terminated that study prematurely, yet a protective effect of folic acid was still observed (ICMR, 2000).

A further double-blind RCT (Kirke et al., 1992) investigated the effect of a 360-μg folic acid supplement taken daily prior to conception and during the first trimester of pregnancy in women with a history of NTDs in the foetus. In this trial there were three

Table 5.1 **Evidence from human intervention trials that folic acid reduces risk of neural tube defects (NTDs)**

Studies	Countries	Subjects	Interventions (μg)	Results (NTD cases/total pregnancies)		
				Intervention groups	Control groups	*P* values
Randomised, controlled trials						
Laurence et al. (1981)	South Wales, United Kingdom	Women prior to conception, previous NTD pregnancy	Folic acid (4000)	0/44	6/67	0.04
MRC (1991)	United Kingdom, Australia, Canada, France, Hungary, Israel, Russia	Women planning a pregnancy, previous NTD pregnancy	Folic acid (4000) or with multivitamin	6/593	21/602	Not reported
Kirke et al. (1992)	Ireland	Women planning a pregnancy, previous NTD pregnancy	Folic acid (360) alone or with multivitamin	0/172	1/89 (*multivitamin control group*)	Not reported
Kirke et al. (1992)	Ireland	Women planning a pregnancy, previous NTD pregnancy	Folic acid (360) alone or with multivitamin	0/172	3/103	Not reported
Czeizel and Dudas (1992), Czeizel et al. (1994)	Hungary	Women planning a pregnancy, no previous NTD pregnancy	Vitamin supplement with folic acid (800) versus trace element supplement	0/2,104	6/2,052	0.029
ICMR (2000)	India	Women planning a pregnancy, previous NTD pregnancy	Vitamin supplement with folic acid (4000)	4/137	10/142	<0.05

Controlled, non-randomised trials						
Smithells et al. (1981)	United Kingdom	Women planning a pregnancy, previous NTD pregnancy	Multivitamin with folic acid (360) >28 days	1/200	13/300	<0.01
Smithells et al. (1983)	United Kingdom	Women planning a pregnancy, previous NTD pregnancy	Multivitamin with folic acid (360) >28 days	2/254	11/219	<0.007
Vergel et al. (1990)	Cuba	Women planning a pregnancy, previous NTD pregnancy	Folic acid (5000)	0/81	4/114	<0.025
Berry et al. (1999)	China	Non-pregnant in area of China with high NTD rates	Folic acid (400)	13/13,012	87/13,369	Not reported
Berry et al. (1999)	China	Non-pregnant in area of China with low NTD rates	Folic acid (400)	34/58,638	86/104,320	Not reported
Czeizel et al. (2004)	Hungary	Women planning a pregnancy, no previous NTD pregnancy	Multivitamin supplement including folic acid (800)	1/3,056	9/3,056	Not reported
Felkner et al. (2005)	USA	Women planning a pregnancy, previous NTD pregnancy	Folic acid supplement (400)	0/75	0/32	Not reported
Felkner et al. (2005)	USA	Women planning a pregnancy, previous NTD pregnancy	Folic acid supplement (4000)	0/86	0/32	Not reported
Non-randomised, non-controlled trials						
Collins et al. (2011)	USA	Non-pregnant, previous NTD pregnancy	Folic acid supplement (4000) if actively trying to conceive, (400) if not trying to conceive	1 Occurrence/418 mothers who took periconceptional folic acid supplements	4 Ocurrences/66 mothers who did not use folic acid supplements	0.0014

treatment groups to which the participants were randomised: folic acid alone, folic acid plus micronutrients and micronutrients alone. There was also a non-randomised control group. There was 1 NTD recurrence in 89 foetuses/infants of women in the multivitamin group and no recurrence in the 172 foetuses/infants in the women taking folic acid. Though these results were not statistically significant, they favoured folic acid intervention [risk ratio (RR) 0.121, 95% CI 0.01–2.29]. However, in comparison with the non-randomised control group in which there were 3 recurrences out of 103 infants, the results were statistically significant.

A Cochrane Systematic Review (De-Regil et al., 2010) has since examined whether folic acid supplementation before and during early pregnancy reduces NTDs without causing adverse outcomes to the mothers and infants. Five trials were included in the analysis (Czeizel et al., 1994; ICMR, 2000; Kirke et al., 1992; Laurence et al., 1981; MRC, 1991). The trials showed a consistent protective effect of daily folic acid supplementation (alone or in combination with other multivitamins) in reducing risk of NTDs compared with placebo, no intervention or micronutrients without folic acid. The overall RR was 0.28, 95% CI 0.00–1.33. The overall effect for recurrence of NTDs was RR 0.32, 95% CI 0.17–0.60. There was no evidence for any short-term side effects of folic acid supplements, and no statistically significant evidence for any effect on miscarriage or any other birth defects.

5.5.2 Observational data

In addition to the evidence from the intervention trials, observational studies in the general population have shown that women taking folic acid supplements have a lower risk of an NTD-affected pregnancy as compared with women not taking folic acid. Studies of dietary intake also indicate that risk of NTDs is reduced with the intake of natural folates from the diet over 230 μg/day. Further relevant data also come from countries with mandatory folic acid fortification policies (such as the USA and Canada) where the target additional intake of 100 μg folic acid/day is associated with a 27% reduction in NTD incidence in the USA and 50% reduction in Canada. These data provide further support for the effect of folic acid in reducing first-time occurrences of NTD.

5.5.3 Evidence relating to the risk factor

The risk factor for the claim is sub-optimal maternal folate status. There are two markers of folate status: serum folate and red blood cell (RBC) folate concentrations. Serum folate is a marker of early changes in folate status and RBC folate is a marker of long-term folate status, reflecting tissue stores of folate. RBCs obtain their folate supply from the serum, concentrating it and then maintaining this level for the lifetime of the cell. RBC folate does not decrease until weeks or even months after reduction in folate intake and the ensuing fall in serum folate concentration. RBC folate thus reflects folate turnover during the previous 120 days.

EFSA concluded that both measures of folate status are informative with regard to the incidence of NTD. The risk factor was taken as an overall maternal folate status rather than either of the measures alone. Measurement of both RBC and serum folate is methodologically valid with respect to analytical characteristics. The same applies for the use of RBC folate as the outcome measure in relation to folic acid intake.

Having established the markers for folate status, it was then necessary to show that folate status influences NTD outcomes and that increasing folic acid intake increases folate status.

A key paper based on a large dataset of NTD cases has clearly established maternal RBC folate as a risk marker for NTD-affected pregnancies (Daly et al., 1995). The data set included 56,049 blood samples collected at a median of 15 weeks gestation from all women attending the first antenatal clinic in Dublin hospitals from 1986 to 1990 (Kirke et al., 1993). Three controls per NTD case were taken from births without NTD of gestations of 23 weeks or more in the same hospital and during the same period as the cases. Blood samples were retrieved for 84 NTD cases (rate = 1.9/1000 live births) and for 266 controls. RBC and plasma folate were measured, and risk of NTDs was estimated in different bands of RBC folate concentrations. RBC folate values were considered a proxy for folate status at the time of closure of the neural tube. A more than eightfold difference in risk was observed between women who had RBC folate levels less than 150 ng/mL (risk of NTDs 6.6/1000 births) compared with women who had RBC levels of 400 ng/mL or above (risk of NTDs 0.8/1000 births) ($P < 0.001$). This demonstrated that RBC folate in early pregnancy is a marker of the risk of NTDs.

It was further shown that the relationship between NTD risk and RBC folate is a continuous dose–response relationship (Daly et al., 1995). The study also demonstrated that the concentration of RBC folate, indicating a risk for NTDs, is higher than the concentration indicating folate deficiency and risk of megaloblastic anaemia (150 ng/mL). Hence NTDs are not maternal folate deficiency diseases as such but relate to folate insufficiency or sub-optimal folate status.

Other studies have investigated the association of NTDs with both serum and/or RBC folate, with most (though not all of the early studies) finding a significant inverse association between serum and/or RBC folate and NTDs. A meta-analysis of studies in which blood samples were taken at antenatal booking or during the first trimester of pregnancy, has confirmed the association between both serum ($P < 0.001$) and RBC folate ($P < 0.001$) and risk of NTDs (Wald et al., 1996). RBC folate was 77 ng/mL lower in cases than in controls. In contrast, in studies with blood samples collected during the second trimester (Economides et al., 1992; Laurence et al., 1981) there was no statistically significant association between RBC folate and risk of NTDs.

RBC and serum folate are strongly correlated, and the meta-analysis by Wald et al. (1996) supports the validity of the association for both measures. The RBC concentration associated with the lowest risk of NTDs is 400 ng/mL or 906 nmol/L (Daly et al., 1995).

The next step in compiling the evidence was to investigate the necessary intake of folic acid to maintain folate status at this optimal concentration. Many studies have demonstrated that folic acid raises maternal RBC folate. In a dose-finding intervention trial, Daly et al. (1997) investigated the dose response to different intakes of folic acid to determine the optimal requirement for women of childbearing age to raise RBC folate concentration to the optimal concentration). In the trial, 121 women with RBC folate concentration between 150 and 400 ng/mL were randomly assigned to daily placebo, 100, 200 or 400 μg folic acid for 6 months. The study duration allowed 8 weeks for folate in the serum to reach a new plateau and for sampling 120 days later when the whole RBC population had been exposed to the higher serum folate concentration. Data analysis was by 'intention to treat', with 95 of the 121 women completing the trial. Post-treatment RBC folate concentration differed significantly among the four groups ($P = 0.001$). RBC folate fell by a median of 12 ng/mL in the placebo group and rose by 67, 130 and 200 ng/mL in the 100, 200 and 400 μg folic acid/day groups respectively. The study demonstrated that 400 μg folic acid/day is the most protective intake, doubling RBC folate concentrations and raising them above the target of 400 ng/mL in all of the compliant women.

Other studies have also shown that folic acid raises serum folate concentrations and Wald et al. (2001) calculated that serum folate concentration increases by 0.94 ng/mL for every 0.1-mg/day increase in folic acid intake in women aged 20–35 years, with every doubling of serum folate concentration halving the risk of NTDs.

In addition to supplementation studies, those investigating the effects of natural folates in the diet and folic acid from fortified foods were also reviewed. While studies have shown that folic acid fortification can raise serum or RBC folate and improve status, studies of dietary folate have not shown such an effect. This inconsistency is explained by the higher bioavailability of folic acid compared with food folates and the greater effectiveness of folic acid supplements or fortified foods in increasing folate status (Cuskelly et al., 1996). In a study investigating predictors of RBC folate in women trying to conceive, daily folate intakes of 450 μg and above in users of folic acid supplements were found to correspond to protective levels of RBC folate (400 ng/mL), whereas non-supplement users required daily intakes of more than 500 μg from food folates and folic acid-fortified cereals to attain protective RBC concentrations. The authors concluded that daily use of a 400-μg folic acid supplement would ensure RBC folate levels above 400 mg/mL in the majority of women (Brown et al., 1997).

It can thus be concluded from the data overall that a 400-μg folic acid supplement per day is the optimal intake of folic acid to raise maternal folate status to the concentration associated with least risk of NTDs.

5.6 Authorisation of the claim

5.6.1 EFSA opinion

The health claim application was submitted in February 2013 to the UK competent authority, and the EFSA opinion was published in November (EFSA, 2013). The EFSA panel on Dietetic Products, Nutrition and Allergies (NDA panel) concluded that a cause-and-effect relationship is established between increasing maternal folate status by supplemental folate intake and reduced risk of NTD, and a positive opinion was published.

5.6.2 Conditions of use for the claim

The EFSA NDA panel concluded that to achieve the claimed effect, 400-μg supplemental folate should be consumed daily for at least 1 month before and up to 3 months after conception by the target group which they defined as women of child-bearing age. These conditions were carried through to the authorised claim.

5.6.3 Proposed claim wording

The EFSA NDA panel concluded that the following wording reflected the scientific evidence:

> *Supplemental folate intake increases maternal folate status. Increasing maternal folate status contributes to the reduction of the risk of NTD. However, the wording was changed during authorisation.*

5.6.4 Authorisation procedures and authorised claim wording

Following publication of the positive EFSA opinion, the regulation authorising the claim was published in October 2014 (EU, 2014). This represents a very quick turnaround for a reduction of disease risk claim, some of which have taken many years to authorise (Sadler, 2015).

During authorisation, there was a discussion of the wording of the claim. Areas of discussion included the need to refer to folic acid in the claim wording, mainly because the pertinent studies substantiating the claim had tested the effect of folic acid supplements rather than dietary sources of folate. Also, as food supplements are labelled as containing 'folic acid', referring to folic acid in the claim would ensure consistency between the name of the product and the claim wording which would benefit consumer understanding.

A further point of discussion was the need to refer to NTDs *in the developing foetus*, again in the interests of consumer understanding. The final authorised wording is:

> *Supplemental folic acid intake increases maternal folate status. Low maternal folate status is a risk factor in the development of neural tube defects in the developing foetus.*

5.6.5 Claim is a special case

It is worthy to note that a disease risk reduction claim for a classical deficiency disease would be unlikely to be authorised, even if it received a positive EFSA opinion. It would likely be concluded that any such claim would contravene Article 3 of the Nutrition and Health Claims Regulation (EU, 2006b). This states that claims must not:

1. Be false, ambiguous or misleading.
2. Give rise to doubt about the safety and/or the nutritional adequacy of other foods.

3. Encourage or condone excess consumption of a food.
4. State, suggest or imply that a balanced and varied diet cannot provide appropriate quantities of nutrients in general. Derogations in the case of nutrients for which sufficient quantities cannot be provided by a balanced and varied diet, including the conditions for their application may be adopted.
5. Refer to changes in bodily functions which could give rise to or exploit fear in the consumer.

However, a reduction of disease risk claim for folic acid in relation to NTDs is a special case for a number of reasons. Firstly, it is widely accepted that the protective effect of folic acid against NTDs is not a classical deficiency disease; the deficiency disease associated with inadequate folate intake is megaloblastic anaemia. Women with an NTD-affected pregnancy are not usually deficient in folate, although they may have lower folate status and higher plasma homocysteine concentrations which is raised when folate status is low. The relationship between folic acid and risk of NTDs thus reflects insufficiency of folate. Secondly, the insufficiency affects only a certain population group (newly pregnant women). Thirdly, in some cases there is also a genetic element to the occurrence of NTDs. This relates to the 677C→T polymorphism of the gene encoding the MTHFR enzyme which converts 5,10 methylenetetrahydrofolate to 5-methyl-THF and provides one-carbon units for metabolism. In people homozygous for the T allele, there is a reduced MTHFR activity, lower serum folate and higher plasma total homocysteine concentration, compared with the 677CC genotype. Meta-analyses have shown that the 677TT genotype is associated with an increased risk of NTD-affected pregnancy. However, provided folate intake is in excess of 600 μg/day, there is no difference in serum folate concentration between these two genotypes. The 677TT variant has high prevalence in some European countries.

As an essential nutrient, folic acid is, therefore, a special case with regard to authorisation of a reduction of disease risk claim.

5.7 Additional data

Further to authorisation of the claim, more recent data from China have shown consistency with the findings that 400 μg folic acid/day raises maternal folate status to the concentration associated with the least risk of NTDs (400 ng/mL or 906 nmol/L). A subsequent analysis shows a sixfold decrease in the estimated risk of NTD with increasing RBC folate from 340 nmol/L compared with 1000 nmol/L (Crider et al., 2014). The study also showed that the higher RBC folate concentration could be achieved with 2–6 months of daily 400-μg folic acid supplementation depending on the starting folate status and relevant genetic influences in the population. This perhaps suggests a need to start supplementation even earlier prior to pregnancy, than currently recommended.

An analysis of blood samples of 1769 adults and 902 children, collected as part of the UK National Diet and Nutrition Survey (NDNS) (PHE, 2015), found that 52.1% of the women of childbearing age (16–49 years) were below the WHO threshold for serum total folate, indicating biochemical folate deficiency (10 nmol/L), and 11.3% were below the threshold of deficiency for RBC folate (340 nmol/L). More young women had concentrations below the RBC threshold than older women (15.6% of 16–24 year olds compared with 10.1% of 35–49 year olds). Overall 75% of the women were below the WHO threshold to minimise risk of having an NTD-affected pregnancy (906 nmol/L; WHO, 2015), the threshold being adjusted to 748 nmol/L due to use of a different method of analysis for the NDNS data compared with that used to determine the WHO criteria. The analysis also revealed that for the overall UK population, mean serum total and RBC folate concentrations are approximately half of those in the USA (assessed by NHANES), post mandatory fortification of bread flour with folic acid.

5.8 Other relevant legislation

The EFSA opinion for the claim states that the health claim applies to all forms of folate authorised for addition to foods and food supplements. Hence, although the claim wording refers to folic acid, any food supplement which uses an authorised source of 'folate' could carry the claim.

Authorised sources are set out in various EU directives and regulations: Annex. II of Regulation (EC) No. 1925/2006 (EU, 2006a), Annex. II of Directive 2002/46/EC (EU, 2002) and Annexes. II and III of Regulation (EC) 1170/2009 (EU, 2009). All forms of folates in these Annexes would be considered as 'folic acid' for the purposes of food supplements. These are pteroylmonoglutamic acid and calcium-L-methylfolate. More recently, Regulation EU 2015/414 amending Directive 2002/46/EC has added (6S)-5-methyltetrahydrofolic acid, glucosamine salt as an approved source (EU, 2015). This is a novel food ingredient that required novel foods approval before being permitted for use.

5.9 How the claim is being used in practice

The claim is in use on product labels of folic acid supplements targeted at women providing 400 μg daily intake. It is also used on websites to reinforce and explain the need to take a folic acid supplement for 1 month before and for the first 12 weeks of pregnancy.

5.10 Consumer issues

5.10.1 Incorporating folic acid into a balanced diet at meaningful levels

It is widely accepted that it is difficult to obtain a daily intake of 400 μg folate from food sources alone, as only few foods are rich in folates (Table 5.2). Natural folates are also unstable and vulnerable to cooking and other losses. However, dietary intake makes a valuable contribution to determining folate status at the start of the childbearing years.

Table 5.2 Food sources of folates

Food items	Folate content (μg/100 g)
Chicken liver, cooked	560
Peanuts, raw	240
Sunflower seeds, dry roasted	237
Spinach, raw	194
Seaweed, dry	188
Chickpeas, boiled	172
Asparagus, boiled	149
Spinach, boiled	146
Endive, raw	142
Lettuce, Romaine	136
Globe artichokes, boiled	119
Broad beans, boiled	104
Okra, boiled	100
Walnuts, dry roasted	91
Avocado	89
Savoy cabbage, raw	80
Wheat bran	79
Sun-dried tomatoes	68
Peas, boiled	63
Camembert cheese	62
Brussels sprouts, boiled	60
Broccoli, boiled	56

Data from USDA, 2016. National Nutrient Databank for Standard Reference. Nutrient Data Laboratory, Beltsville, USA.

Ingestion of a set amount of highly bioavailable folic acid in the form of a dietary supplement is the most reliable way for the target group of the claim to feel confident that they are consuming sufficient additional folic acid to raise folate status and hence to reduce the risk of NTD in the foetus. As the requirement for folic acid to protect against NTD represents a relatively high intake, a policy of supplementation targeted at the vulnerable group is a key risk reduction policy.

5.10.2 *Recommendations for supplemental folic acid*

Within the European Union, 17 Member States have either given guidance reflecting the need for folic acid supplementation before conception or have established such a recommendation, including Austria, Belgium, Denmark, Estonia, Finland, France, Germany, Hungary, Ireland, Italy, Luxembourg, The Netherlands, Poland, Portugal, Spain, Sweden and the United Kingdom (Joint Report on NTD, 2011). For example in the United Kingdom, to prevent the first occurrence of NTD, the DH recommends that all women who could become pregnant should take a daily supplement of 400 μg folic acid prior to conception and during the first 12 weeks of pregnancy (DH, 1992). This measure would be expected to reduce NTD-affected pregnancies by 70%. For women with a previous pregnancy affected by an NTD or with a family history of spina bifida, the recommendation is an extra 5 mg folic acid/day which is prescribed as a medicine (SACN, 2006). These daily intakes of folic acid are in addition to the reference nutrient intake of folic acid for pregnancy (300 μg folate/day in the United Kingdom). Women who could become pregnant are also advised to eat more foods naturally rich in folate and foods fortified with folic acid, especially breakfast cereals (DH, 2000).

This advice is in line with the Standard for the prevention of NTDs developed by the Department of Making Pregnancy Safer at WHO (WHO, 2006), which recommends that women trying to conceive take 400 μg folic acid/day prior to conception until 12 weeks of pregnancy, and that women with a previously affected pregnancy take 5 mg folic acid/day and increase their intake of dietary folate.

A study comparing the guidelines across Europe found that of 35 guidelines in 20 European countries (including countries outside the European Union), the recommendations vary. The main points of discrepancy relate to the time period to start taking folic acid before conception (ranging from no specific timeline given, to up to 4 weeks prior to conception). The target population also varies from women planning a pregnancy to all women of childbearing age or to all women who might become pregnant. There are also variations in the higher dose recommended for women with a previous NTD-affected pregnancy (Cawley et al., 2016).

However, despite this public health advice being in place for many years, the uptake of supplements by the target group remains low. Research in the United Kingdom found that only 44% of 238 women in the study who became pregnant had taken any folic acid supplements in the 3 months prior to pregnancy, and only 5.5% had taken the requisite 400 μg daily intake. Other studies have also generally found that less than half of studied women take any folic acid prior to pregnancy (Inskip et al., 2009). A study in Northern Ireland found that of 226 women who reported taking 400-μg folic acid supplements in the first trimester of pregnancy, only 57 (19%) had started to take it before conception and continued until the end of the first trimester of pregnancy, the majority having started folic acid later than the recommended trimeframe (McNulty et al., 2011). This study also showed that starting to take folic acid 6 or more weeks after conception was associated with lower RBC folate concentration compared with starting to take it before conception or at weeks 0–6.

5.10.3 Consumer understanding

It is not clear if low compliance with folic acid supplementation recommendations reflects a lack of understanding of the message, lack of awareness or lack of motivation because women don't think it applies to them at that particular time. A study investigating knowledge and preconceptional use of folic acid in ethnic groups in the United Kingdom found that Caucasians are more likely to take folic acid supplements before conceiving than non-Caucasians, and less knowledge of the importance of folic acid was associated with lower folic acid usage (Peake et al., 2013). A study in Japan has found that awareness of the role of folic acid in protecting against NTDs promoted an increase in supplement use (Kondo et al., 2015).

There have been calls for more sustained public awareness campaigns to raise awareness and understanding of the message (FSAI, 2016; Kondo et al., 2015). Even if mandatory fortification were to be introduced, the recommendation to take a daily 400-μg folic acid supplement would remain in place, as the background level of fortification would not provide sufficient additional folic acid intake for women to be fully protected. In this case it would be even more important to communicate the need to take a daily supplement to the target group of women, as they may mistakenly believe that folic acid fortification is sufficient to protect them.

5.11 Future trends: mandatory fortification with folic acid?

To reduce the prevalence of NTD, other countries, including the United States, Canada, Chile, South Africa, Argentina, Brazil, and Costa Rica amongst others, have introduced mandatory folic acid fortification of various grain products, for example flour. In the USA, as the 1992 recommendation for women who could become pregnant to take a daily 400-μg supplement, followed by a major public health campaign, resulted in limited uptake of the advice, and considering that an estimated 50% of pregnancies are unplanned in the USA, the US Food and Drug Administration authorised mandatory fortification. This was for the addition of 140 μg folic acid/100 g of grain of all enriched cereals and grains, with compliance mandatory by January 1998. Health Canada endorsed a similar fortification strategy in 1998 mandating 0.15 mg/100 g of grain in white flour, pasta and cornmeal. In Chile the mandatory fortification of wheat flour was introduced in 2000, with 220 μg folic acid/100 g flour. These countries have reported reductions in the prevalence of NTD-affected births of 19%–55%, indicating that mandatory fortification is an effective public health measure.

Ireland has one of the highest global rates of NTD. It has been estimated that only 36% of women of childbearing age in Ireland have RBC folate status sufficient for optimal protection against NTD. Women of childbearing age have been advised to take a daily 400-μg folic acid supplement but the uptake is low. Voluntary fortification introduced in the early 1980s has helped to reduce the risk of NTD-affected pregnancies, but this has been less effective than mandatory fortification policies seen in countries such as the USA. The Food Safety Authority of Ireland has proposed two policy options for the future: mandatory fortification with continued voluntary fortification and advice on daily 400-μg folic acid supplementation or continuing with voluntary fortification and advice on supplementation. Mandatory fortification would provide approximately 150 μg folic acid/day in women of childbearing age which is estimated to reduce the prevalence of NTDs by about 30% (FSAI, 2016).

In the United Kingdom the Committee on Medical Aspects of Food Policy (COMA) explored five possible levels of folic acid fortification of flour and concluded that 240 μg/100 g would have a significant effect in reducing NTD-affected pregnancies without resulting in unacceptable high intakes in any group of the population. However, education to maintain high levels of awareness and advice to take a daily 400-μg supplement would need to continue (DH, 2000). Mandatory fortification was not introduced at that time due to concerns about masking vitamin B12 deficiency in the older population which might allow the associated neuropathy to develop unchecked. A further modelling exercise undertaken for the Scientific Committee on Nutrition report (SACN, 2006) identified that mandatory fortification would reduce the risk of NTDs but would increase the number of people consuming above the upper level (UL) of folic acid (1 mg/day) and the number of older people with low vitamin B12 status consuming above 1 mg folic acid/day. Hence it was recommended that mandatory fortification should be accompanied by controls on voluntary fortification to minimise the number of people consuming above the safe level, in which case a level of 300 μg/100 g flour would be optimal and would reduce risk of NTDs by 11%–18% (approximately 160 NTD-affected pregnancies per year). Advice to take a daily supplement of 400 μg folic acid would still be necessary. A further statement from SACN (2009) recommended that guidance on folic acid supplement use for particular population groups would also be necessary. The DH had discussions with industry at the time, to consider whether it was appropriate to limit voluntary fortification of foods and the folic acid content of supplements not targeted at women of childbearing age. However, the UK government announced in 2016 that it would not go ahead with mandatory fortification with folic acid.

Clearly, with mandatory fortification policies the challenge is to raise the intake of folic acid in the target group of women of childbearing age but not to increase the number of adults or children who exceed the UL (Martiniak et al., 2015). The UL in Europe for folic acid is set at 1 mg/day for adults including pregnant and lactating women (EFSA, 2006). The UL has an important bearing on the potential level of fortification and on policy for voluntary fortification of foods, particularly in consideration of the non-target population groups that would be exposed to additional folic acid, and other population groups taking folic acid supplements. Concerns have been raised that increasing folic acid intake across entire populations may increase risk of adverse effects for some of the non-target groups.

For example concerns have been raised about the interpretation of the early evidence that folic acid masks B12 deficiency, thus allowing the associated neuropathy to progress undetected (Reynolds, 2016). Reynolds argues that, rather than just masking deficiency, the early data suggest that folic acid may progress or precipitate the neurological manifestations of anaemia and that the UL should be set at a lower intake. This is mainly of relevance to older people, as ageing is associated with a decline in vitamin B12 status. In the face of increasing calls for mandatory and voluntary fortification, Reynolds has proposed an interesting solution to the dilemma of potential increasing folic acid intakes and the neurological risks in older people. He suggests that mandatory folic acid fortification policies should be accompanied by vitamin B12 fortification. As there is evidence that vitamin B12 deficiency is also a risk factor for NTDs, addition of vitamin B12 should further reduce any risk from high-folic acid intakes resulting from fortification policies.

Concerns have also been raised that population-wide exposure to increased intakes of folic acid may increase risk of colorectal cancer. However, after examining the data, SACN in the United Kingdom upheld their previous recommendation for mandatory fortification (SACN, 2009). Another concern is the potential for increased concentrations of free (unmetabolised) folic acid in the blood, the effects of which are unknown. There are currently no established thresholds for free folic acid and limited population data on concentrations.

Mandatory fortification also raises several technical, trade policy and food-labelling issues that would need to be addressed (Richardson, 2015), for example the issue of overages needed to counter losses during baking and to allow for shelf life, the need to label fortificants in flour and implications for the import and export of flours. Despite exploration of mandatory fortification policies, for example in the United Kingdom, Germany, Ireland and Finland, currently no EU Member State has initiated a mandatory fortification policy, the United Kingdom having since confirmed that no such policy will be introduced in the near future.

5.12 EU DRVs for folate intake

In 2014, EFSA set dietary reference values for folate (EFSA, 2014). The average requirement was determined from the intake required to achieve a serum folate concentration of ≥10 nmol/L and an RBC folate concentration of 340 nmol/L, reflecting maintenance of folate adequacy. For adults, an average daily requirement of 250 μg DFE and a population reference intake of 330 μg/day was set. For pregnant women, an adequate intake of 600 μg/day was derived on the basis of maintaining adequate serum and RBC folate levels. The requirement for 400-μg folic acid supplementation prior to and for the first trimester of pregnancy is in addition to the 600 μg required for pregnancy. It will be difficult to achieve the intake of 600 μg required for pregnancy from food folates alone, such that the requirement for supplementation during the first trimester of pregnancy is ideally higher than the 400 μg needed for protection against NTDs. However, this would make a more complicated message to communicate to women.

5.13 Conclusions

The need to take a daily 400-μg folic acid supplement, starting at least 1 month prior to conception and during the first trimester of pregnancy to reduce the risk of NTDs in the foetus, is long established and is a global WHO recommendation. As the number of NTD-affected pregnancies remains high in Europe, there is an ongoing need to ensure that women of childbearing age are aware of this recommendation. Allowing a reduction of disease risk health claim on supplements containing 400 μg folic acid/day provides a new opportunity to reinforce this message on packaging and via websites and in marketing and advertising.

5.14 Sources of further information and advice

EFSA, 2016. General scientific guidance for stakeholders on health claim applications. EFSA J. 14 (1), 4367.

The Commission website (http://ec.europa.eu/nuhclaims) includes the register of EU health claims.

The European Parliament and the Council, 2006. Corrigendum to Regulation (EC) No. 1924/2006 of the European Parliament and of the Council of 20 December 22006 on nutrition and health claims made on foods. Off. J. Eur. Union L 12/3.

References

Berry, R.J., Li, Z., Erickson, J.D., et al., 1999. Prevention of neural-tube defects with folic acid in China. N. Engl. J. Med. 341 (20), 1485–1490.

Brown, J.E., Jacobs, Jr., D.R., Hartman, T.J., et al., 1997. Predictors of red cell folate level in women attempting pregnancy. JAMA 277, 548–552.

Caudill, M.A., 2010. Folate bioavailability: implications for establishing dietary recommendations and optimizing status. Am. J. Clin. Nutr. 91, 1455S–1460S.

Cawley, S., Mullaney, L., McKeating, A., et al., 2016. A review of European guidelines on periconceptional folic acid supplementation. Eur. J. Clin. Nutr. 70, 143–154.

Collins, J.S., Atkinson, K.K., Dean, J.H., et al., 2011. Long term maintenance of neural tube defects prevention in a high prevalence state. J. Pediatr. 159 (1), 143–149.

Crider, K.S., Devine, O., Hao, L., et al., 2014. Population red blood cell folate concentrations for prevention of neural tube defects: Bayesian model. Br. Med. J. 349, g4554.

Cuskelly, G.J., McNulty, H., Scott, J.M., 1996. Effect of increasing dietary folate on red-cell folate: implications for prevention of neural tube defects. Lancet 347, 657–659.

Czeizel, A.E., Dobo, M., Vargha, P., 2004. Hungarian cohort-controlled trial of periconceptional multivitamins supplementation shows a reduction in certain congenital abnormalities. Birth Defects Res. A 70, 853–861.

Czeizel, A.E., Dudas, I., 1992. Prevention of the first occurrence of neural-tube defects by periconceptional vitamin supplementation. N. Engl. J. Med. 327, 1832–1835.

Czeizel, A.E., Dudas, I., Metneki, J., 1994. Pregnancy outcomes in a randomised controlled trial of periconceptional multivitamin supplementation. Final report. Arch. Gynecol. Obstet. 255, 131–139.

Daly, L.E., Kirke, P.N., Molloy, A., et al., 1995. Folate levels and neural tube defects. Implications for prevention. JAMA 274, 1698–1702.

Daly, S., Mills, J.L., Molloy, A.M., et al., 1997. Minimum effective dose of folic acid for food fortification to prevent neural-tube defects. Lancet 350, 1666–1669.

Department of Health (DH), 1992. Folic acid and the prevention of neural tube defects. Report from an Expert Advisory Group (Available from DH, Wetherby).

Department of Health (DH), 2000. Folic acid and the prevention of disease. Report of the Committee of the Medical Aspects of Food and Nutrition Policy, No. 50. HMSO, London.

De-Regil, L.M., Fernández-Gaxiola, A.C., Dowswell, T., Pena-Rosas, J.P., 2010. Effects and safety of periconceptional folate supplementation for preventing birth defects. Cochrane Database of Syst. Rev. (10), CD007950.

EC, 2008. European Commission Communication on Rare Diseases: Europe's Challenges (COM(2008) 679), November 2008. Available from: http://ec.europa.eu/health/ph_threats/non_com/docs/rare_com_en.pdf.

Economides, D.L., Ferguson, J., Mackenzie, I.Z., et al., 1992. Folate and vitamin B12 concentrations in maternal and fetal blood, and amniotic fluid in second trimester pregnancies complicated by neural tube defects. Br. J. Obstet. Gynaecol. 99 (1), 23–25.

EFSA, 2006. Tolerable Upper Intake Levels for Vitamins and Minerals. EFSA, Parma.

EFSA, 2009a. Scientific opinion on the substantiation of a health claim related to calcium plus vitamin D3 chewing tablets and reduction of the risk of osteoporotic fractures by reducing bone loss pursuant to Article 14 of Regulation (EC) No. 1924/2006. EFSA J. 1180, 1–13.

EFSA, 2009b. Scientific opinion on the substantiation of health claims related to folate pursuant to Article 13(1) of Regulation (EC) No. 1924/2006. EFSA J. 7 (9), 1213.

EFSA, 2011. Scientific opinion on the substantiation of a health claim related to vitamin D and risk of falling pursuant to Article 14 of Regulation (EC) No. 1924/2006. EFSA J. 9 (9), 2382.

EFSA, 2013. Scientific opinion on the substantiation of a health claim related to increasing maternal folate status by supplemental folate intake and reduced risk of neural tube defects pursuant to Article 14 of Regulation (EC) No. 1924/2006. EFSA J. 11 (7), 3328.

EFSA, 2014. Scientific opinion on dietary reference values for folate. EFSA J. 12 (11), 3893.

EU, 2002. Directive 2002/46/EC of the European Parliament and of Council of 10 June 2002 on the approximation of the laws of the Member States relating to food supplements. Off. J. Eur. Commun. L 183/51.

EU, 2006. Regulation (EC) No. 1925/2006 of 20 December 2006 on the addition of vitamins and minerals and of certain other substances to foods. Off. J. Eur. Commun. L 404/26.

EU, 2006. Regulation (EC) No. 1924/2006 of the European Parliament and of the Council of 20 December 2006 on nutrition and health claims made on foods. Off. J. Eur. Commun. 404.

EU, 2009. Regulation (EC) 1170/2009 of 30 November 2009 amending Directive 2002/46/EC of the European Parliament and of Council, and Regulation (EC) 1925/2006 of the European Parliament and of Council, as regards the lists of vitamin and minerals and their forms that can be added to foods including food supplements. Off. J. Eur. Commun. L 314/36.

EU, 2012. Commission Regulation (EU) No. 432/2012 of 16 May 2012 establishing a list of permitted health claims made on foods, other than those referring to the reduction of disease risk and to children's development and health. Off. J. Eur. Commun. L 131.

EU, 2014. Commission Regulation (EU) No. 1135/2014 of 24 October 2014 on the authorization of a health claim made on foods and referring to the reduction of disease risk. Off. J. Eur. Commun. L 307/23.

EU, 2015. Commission Regulation EU 2015/414 of 12 March 2015 amending Directive 2002/46/E as regards (6S)-5-methyltetrahydrofolic acid, glucosamine salt used in the manufacture of food supplements. Off. J. Eur. Commun. L 68/26.

Felkner, M., Suarez, L., Hendricks, K., Larsen, R., 2005. Implementation and outcomes of recommended folic acid supplementation in Mexican-American women with prior neural tube defect-affected pregnancies. Prev. Med. 40, 867–871.

FSAI, 2016. Update on Folic Acid and the Prevention of Birth Defects in Ireland. Report of the Scientific Committee of the Food Safety Authority of Ireland. FSAI, Dublin.

Indian Council of Medical Research (ICMR) Collaborating Centres and Central Technical Coordinating Unit, ICMR, 2000. Multicentric study of efficacy of periconceptional folic acid containing vitamin supplementation in prevention of open neural tube defects from India. Indian J. Med. Res. 112, 206–211.

Inskip, H.M., Crozier, S.R., Godfrey, K.M., et al., 2009. Women's compliance with nutrition and lifestyle recommendations before pregnancy: general population cohort study. Br. Med. J. 338, b481.

Joint Report on NTD, 2011. Act against Europe's most common birth defects: one year on—defining neural tube defect prevention strategies in Europe.

Kirke, P.N., Daly, L.E., Elwood, J.H., For the Irish Vitamin Study Group, 1992. A randomised trial of low dose folic acid to prevent neural tube defects. Arch. Dis. Child. 67, 1442–1446.

Kirke, P.N., Molloy, A.M., Daly, L.E., et al., 1993. Maternal plasma folate and vitamin B12 are independent risk factors for neural tube defects. Q. J. Med. 86 (11), 703–708.

Kondo, A., Morota, N., Date, H., et al., 2015. Awareness of folic acid use increases its consumption, and reduces risk of spina bifida. Br. J. Nutr. 114, 84–90.

Laurence, K.M., James, N., Miller, M.H., et al., 1981. Double-blind randomised controlled trial of folate treatment before conception to prevent recurrence of neural-tube defects. Br. Med. J. 282, 1509–1511.

Martiniak, Y., Heuer, T., Hoffmann, I., 2015. Intake of dietary folate and folic acid in Germany based on different scenarios for food fortification with folic acid. Eur. J. Nutr. 54, 1045–1054.

McNulty, H., Pentieva, K., 2004. Folate bioavailability. Proc. Nutr. Soc. 63, 529–536.

McNulty, B., Pentieva, K., Marshall, B., et al., 2011. Women's compliance with current folic acid recommendations and achievement of optimal vitamin status for preventing neural tube defects. Hum. Reprod. 26, 1530–1536.

Medical Research Council (MRC) Vitamin Study Research Group, 1991. Prevention of neural tube defects: results of the medical research council vitamin study. Lancet 338, 131–137.

Peake, J.N., Copp, A.J., Shawe, J., 2013. Knowledge and periconceptional use of folic acid for the prevention of neural tube defects in ethnic communities in the United Kingdom: systematic review and meta-analysis. Birth Defects Res. A 97, 444–451.

PHE, 2015. National Diet and Nutrition Survey Rolling Programme (NDNS RP) Supplementary Report: Blood Folate Results for the UK as a Whole, Scotland, Northern Ireland (Years 1 to 4 combined) and Wales (Years 2 to 5 Combined). PHE, London.

Reynolds, E.H., 2016. What is the safe upper intake level of folic acid for the nervous system? Implications for folic acid fortification policies. Eur. J. Clin. Nutr. 70, 537–540.

Richardson, D.P., 2015. Developing the right public health strategies for folic acid and reduction of risk of neural tube defects (NTDs) in the United Kingdom. Eur. J. Nutr. Food Saf. 5 (4), 242–249.

SACN, 2006. Folate and Disease Prevention. TSO, London.

SACN, 2009. Folic Acid and Colorectal Cancer Risk: Review of Recommendations for Mandatory Folic Acid Fortification. PHE, London.

Sadler, M.J., 2015. Regulatory developments with European health claims. Sadler, M.J. (Ed.), Foods, Nutrients and Food Ingredients with Authorsied EU Health Claims, vol. 2, Woodhead Publishing, Cambridge, United Kingdom.

Smithells, R.W., Nevin, N.C., Seller, M.J., et al., 1983. Further experience of vitamin supplementation for prevention of neural tube defect recurrences. Lancet 1, 1027–1031.

Smithells, R.W., Sheppard, S., Schorah, C.J., et al., 1981. Apparent prevention of neural tube defects by periconceptional vitamin supplementation. Arch. Dis. Child. 56, 911–918.

Vergel, R.G., Sanchey, L.R., Heredero, B.L., et al., 1990. Primary prevention of neural tube defects with folic acid supplementation: Cuban experience. Prenat. Diag. 10, 149–152.

Wald, N.J., Hackshaw, A.D., Stone, R., Sourial, N.A., 1996. Blood folic acid and vitamin B12 in relation to neural tube defects. Br. J. Obstet. Gynaecol. 103, 319–324.

Wald, N.J., Law, M.R., Morris, J.K., Wald, D.S., 2001. Quantifying the effect of folic acid. Lancet 358, 2069–2073.

WHO, 2006. Standards for Maternal and Neonatal Care, Prevention of Neural Tube Defects. Department of Making Pregnancy Safer, World Health Organization, Geneva.

WHO, 2012. Guideline: Daily Iron and Folic Acid Supplementation in Pregnant Women. World Health Organization, Geneva.

WHO, 2015. Guideline: Optimal Serum and Red Blood Cell Folate Concentrations in Women of Reproductive Age for Prevention of Neural Tube Defects. World Health Organization, Geneva.

Winkels, R.M., Brouwer, I.A., Siebelink, E., et al., 2007. Bioavailability of food folates is 80% of that of folic acid. Am. J. Clin. Nutr. 85, 465–473.

Authorised EU health claim for MUFA and PUFA in replacement of saturated fats

6

Katherine M. Livingstone
Institute for Physical Activity and Nutrition, Deakin University, Geelong, VIC, Australia

6.1 Introduction

The following chapter will evaluate the disease risk reduction health claim application that was related to low-fat and low-*trans* spreadable fat rich in unsaturated and omega-3 fatty acids and reduction of low-density lipoprotein (LDL)-cholesterol concentrations (EFSA Panel on Dietetic Products Nutrition and Allergies, 2011a). The characterisation of the food constituents, the scientific substantiation for the health claim and the conditions of use will be defined and evaluated. Moreover, the wider impact of this claim will be discussed in relation to potential consequences for consumers, the development of products and future trends.

6.2 Characterisation of the substance

The constituents subject to characterisation were unsaturated fatty acids (UFA), including *trans* fatty acids (TFA) as well as saturated fatty acids (SFA). This claim referred specifically to replacing mixtures of SFAs and TFAs present in foods or diets with mixtures of UFAs [mixtures of *cis*-monounsaturated fatty acids (MUFAs) and/or *cis*-polyunsaturated fatty acids (PUFAs)]. EFSA concluded that these constituents were sufficiently characterised and the EFSA definitions for UFAs and SFAs are quoted in the following sections.

6.2.1 Unsaturated fatty acids

> Cis-*monounsaturated fatty acids (MUFAs) have one double bond in the fatty acid (FA) chain and the quantitatively most important representative in the diet and in tissue lipids is oleic acid [18:1* (n−9)*].* Cis-*polyunsaturated fatty acids (PUFAs) have two to six double bonds in the FA chain and the most abundant* n−6 *and* n−3 *PUFAs in the diet are linoleic acid (LA, 18:2* (n−6)*) and alpha-linolenic acid [ALA, 18:3* (n−3)*] respectively. The long-chain PUFAs, with chain length greater than 18C, are not considered in this opinion. Most UFAs (i.e. MUFA and PUFA) in the diet have the* cis *configuration, but TFAs are also present. These FAs originate from several sources and* trans-*MUFAs are the most common TFA in the*

Foods, Nutrients and Food Ingredients with Authorised EU Health Claims. http://dx.doi.org/10.1016/B978-0-08-100922-2.00006-1
Copyright © 2018 Elsevier Ltd. All rights reserved.

diet. Trans-*PUFAs, however, are also present.* Trans-*PUFAs have at least one* trans *double bond and may therefore also have double bonds in the* cis *configuration.*
(EFSA Panel on Dietetic Products Nutrition and Allergies, 2011a).

6.2.2 Saturated fatty acids

SFAs are aliphatic monocarboxylic acids with (generally) an even number of carbon atoms (usually from 4 to 20) and no double bonds, that can be liberated by hydrolysis of triacylglycerols from fats and oils. The most prevailing SFAs in the diet are lauric acid (12:0), myristic acid (14:0), palmitic acid (16:0) and stearic acid (18:0).
(EFSA Panel on Dietetic Products Nutrition and Allergies, 2011a).

6.3 Authorised EU health claim: UFA and LDL-cholesterol

The EFSA Panel on Dietetic Products, Nutrition and Allergies (NDA) evaluated the scientific substantiation for the health claim relating to 'low fat and low *trans* spreadable fat rich in unsaturated and omega-3 fatty acids' and reduction of LDL-cholesterol concentrations. This claim is under Article 14.1 (a) of the EC Regulation on nutrition and health claims, thereby denoting a disease risk reduction claim. In this respect, the target population is people who want to lower their blood cholesterol. The EFSA panel gave a positive opinion, suggesting the following claim wording and conditions of use.

6.3.1 Claim wording

The proposed wording, as defined by the EFSA Panel was as follows: 'Consumption of saturated fat increases blood cholesterol concentrations; consumption of mono- and/or polyunsaturated fat in replacement of saturated fat has been shown to lower/reduce blood cholesterol. Blood cholesterol lowering may reduce the risk of (coronary) heart disease' (EFSA Panel on Dietetic Products Nutrition and Allergies, 2011a).

The final authorised wording approved for use in the EU was as follows: 'Replacing saturated fats with unsaturated fats in the diet has been shown to lower/reduce blood cholesterol. High cholesterol is a risk factor in the development of coronary heart disease' (EFSA Panel on Dietetic Products Nutrition and Allergies, 2011a). This wording is consistent with other authorised disease risk reduction claims for cholesterol reduction.

6.3.2 Conditions of use

To bear the claim, the EFSA Panel specified that large proportions of mixed SFAs in foods or diets should be replaced on a gram-per-gram basis by *cis*-MUFAs and/or *cis*-PUFAs. These conditions are in accordance with the Annex of Regulation (EC) No. 1924/2006 (EU Regulation (EC), 2006), its recent amendment Regulation (EC) No. 116/2010 (EU Regulation (EC), 2010), and the Guidance on the implementation of Regulation (EC) No. 1924/2006 of the Standing Committee on the Food Chain and Animal Health for comparative nutrition claims made on foods (European Commission, 2007).

The given claim relates to foods with a high MUFA and PUFA content. As such, the conditions of the nutrition claim 'High Unsaturated Fat', as listed in the Annex to Regulation (EC) No. 116/2010, should be met to make this health claim (EU Regulation (EC), 2006). These conditions are as follows: 'At least 70% of the FAs present in the product derive from unsaturated fat under the condition that unsaturated fat provides more than 20% of energy of the product'. Furthermore, the claim may only be used on fats and oils, and the target population is people who want to lower their blood cholesterol (EU Regulation (EC), 2014).

6.3.3 Summary of substantiating evidence

A recent scientific opinion, published by the EFSA NDA Panel, evaluated the scientific substantiation for the health claim relating to replacement of mixtures of SFAs with *cis*-MUFAs and/or *cis*-PUFAs in foods or diets and *maintenance* of normal blood LDL-cholesterol concentrations (EFSA Panel on Dietetic Products Nutrition and Allergies, 2011c). This opinion highlighted that the totality of the evidence was strong enough to support this relationship, which facilitated an opinion by the EFSA Panel that this substantiation would also apply to the *reduction* of blood LDL-cholesterol concentrations, thus supporting the present claim. Given the specific mention of TFAs in the EFSA scientific opinion, the Panel reviewed the evidence in relation to TFAs and LDL cholesterol. The following section will summarise and evaluate the main sources of scientific evidence used to substantiate the claimed effect between UFA and LDL-cholesterol concentrations (Table 6.1). The full list of supporting references provided to EFSA is available on the EFSA website (EFSA, 2015).

6.3.3.1 EFSA opinions

As reviewed in an appraisal of the substantiation of the EU health claim on replacement of mixtures of SFAs in foods or diets with mixtures of MUFAs and/or PUFAs and maintenance of normal blood LDL-cholesterol concentrations (Sadler, 2015), there are a number of EFSA opinions that have addressed the role of SFAs and TFAs in increasing blood total and LDL-cholesterol. The effects of these FAs on health were reviewed in a 2004 publication by the EFSA Panel (EFSA Panel on Dietetic Products Nutrition and Allergies, 2004), where a meta-analysis by Mensink et al. (2003) was the primary substantiating evidence for a detrimental effect of TFAs on LDL-cholesterol. As reported by Mensink et al. (2003), when 1% of energy in the diet from carbohydrates was replaced isoenergetically by TFAs, LDL-cholesterol increased by 0.040 mmol/L. Similarly, individual SFAs also increased LDL-cholesterol concentrations: palmitic acid (+0.039 mmol/L), lauric (+0.052 mmol/L) and myristic (+0.048 mmol/L). As a result, the EFSA Panel published a scientific opinion on the establishment of dietary reference values for fats (EFSA Panel on Dietetic Products Nutrition and Allergies, 2010), which specified that SFA and TFA intakes should be as low as possible. In contrast, no dietary reference values were set for *cis*-MUFA, as they are indispensable components of the diet, and similarly none were set for *cis*-PUFA given the differential metabolic effects of the various *cis*-PUFA.

Table 6.1 **Details of studies and reports in support of the ESFA claim**

References	Types	Main outcomes	Supporting references
EFSA (2010)	Scientific Opinion	Consumption of TFA, as with SFA, increase LDL-C, compared with consumption of *cis*-MUFA and/or PUFA	Mensink et al. (2003)
EFSA Panel on Dietetic Products Nutrition and Allergies (2010)	Scientific Opinion	1. SFA and TFA intake should be as low as possible. 2. No dietary reference value for *cis*-MUFA intake 3. No dietary reference value for *cis*-PUFA intake 4. No specific values for the ratio of PUFA n−3/n−6	Mensink et al. (2003),
Institute of Medicine (2005)	Dietary advice	Dietary reference intakes: 1. Limit SFA and TFA intake to as low as possible 2. Consume between 20 and 35% energy from $n-6$ PUFA and 5%–10% energy from $n-3$ PUFA	NA
Mensink et al. (2003)	Meta-analysis of clinical trials	1. A 1% replacement of energy from carbohydrates with *trans*-MUFA increases the total cholesterol: LDL-C ratio, while with SFA this ratio marginally increases and with *cis*-MUFA and *cis*-PUFA it significantly reduces 2. Not all SFA have the same cholesterol-raising effect; those with a carbon chain length between 12 and 16 increase LDL-C to a greater extent	60 Randomised trials
Mozaffarian and Clarke (2009)	Meta-analysis of clinical trials	1. A 1% energy replacement of TFAs with SFAs, MUFAs or PUFAs respectively, decreased the total cholesterol: HDL-C ratio by 0.31, 0.54 and 0.67 2. In prospective cohort studies, each 2% energy replacement of TFAs with SFAs, MUFAs or PUFAs lowered CHD risk by 17, 21 and 24% respectively	13 Randomised trials
Lichtenstein et al. (2006)	Dietary advice	1. 'Aim for a desirable lipid profile' <100 mg/dL LDL-C 2. 'Limit intake of SFA, TFA and cholesterol': <7% energy from SFA, typically by replacing with MUFA and/or PUFA	Lichtenstein et al. (1999)
WHO/FAO (2003)	Dietary advice	There is convincing evidence that SFA and TFA increase risk of CVD via their effects on LDL-C	Mensink and Katan (1992)

CVD, Cardiovascular disease; HDL-C, high-density lipoprotein; LDL-C, low-density lipoprotein; MUFA, mono-unsaturated fatty acids; NA, not available; PUFA, poly-unsaturated fatty acids; SFA, saturated fatty acids; TFA, *trans* fatty acids.

6.3.3.2 Randomised controlled trials

A large proportion of the scientific substantiation for the present claim is based on meta-analyses of randomised controlled trials (RCTs). Based on data from 60 trials and 1672 individuals, Mensink et al. (2003) calculated changes in total, LDL- and HDL-cholesterol and the total to HDL-cholesterol ratio. Comparison of replacement of 1% energy from carbohydrates with SFA, *cis*-MUFA, *trans*-MUFA and *cis*-PUFA (excluding long-chain $n-3$ PUFA) showed that the ratio of total to HDL-cholesterol did not change when SFA replaced carbohydrates. Moreover, when replacing *trans*-MUFA with *cis*-MUFA and *cis*-PUFA, Mensink et al. (2003) showed that the ratio decreased by twice as much as when they replaced SFA (Fig. 6.1).

A subsequent meta-analysis by Mozaffarian and Clarke (2009) evaluated evidence for the effect of replacing TFAs with SFAs and UFAs from both RCTs (13 studies included) and observational studies (4 studies included). From the trial data, each 1% energy replacement of TFAs with SFAs, MUFAs or PUFAs decreased the total to HDL-cholesterol ratio by 0.31, 0.54 and 0.67 respectively (Fig. 6.2). Similarly, in prospective cohort studies, each 2% energy replacement of TFAs with SFAs, MUFAs or PUFAs lowered coronary heart disease (CHD) risk by 17, 21 and 24% respectively.

Although both meta-analyses provide evidence of a greater detrimental effect of TFA on blood cholesterol than SFAs, the Panel considered there was insufficient evidence to conclude that there is a meaningful difference between them on a gram-per-gram basis. This conclusion may be in part due to the lack of evidence surrounding ruminant TFAs, for which the association with blood cholesterol concentrations is not as strong as for industrially produced TFAs. Moreover, the Panel noted that even

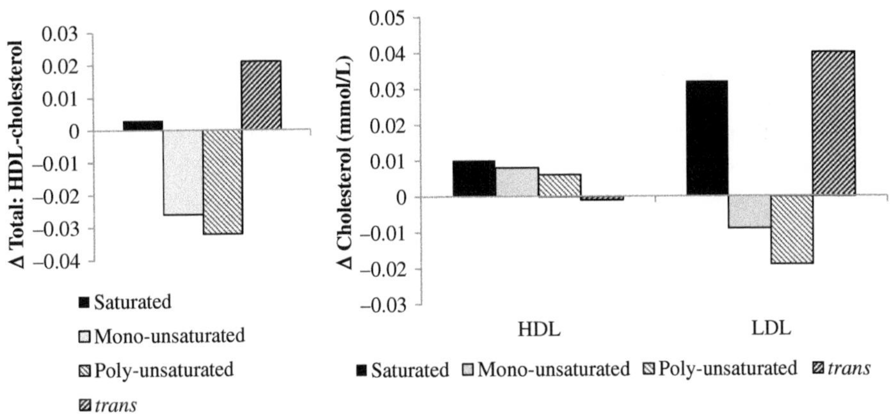

Figure 6.1 Predicted changes (Δ) in the ratio of serum total to high-density lipoprotein (HDL)-cholesterol and in low-density lipids (LDL)- and HDL-cholesterol concentrations when carbohydrates constituting 1% of energy are replaced isoenergetically with saturated, *cis*-monounsaturated, *cis*-polyunsaturated or *trans*-monounsaturated fatty acids.
Source: Mensink, R.P., Zock, P.L., Kester, A.D., Katan, M.B.,2003. Effects of dietary fatty acids and carbohydrates on the ratio of serum total to HDL-cholesterol and on serum lipids and apolipoproteins: a meta-analysis of 60 controlled trials. Am. J. Clin. Nutr. 77, 1146–1155.

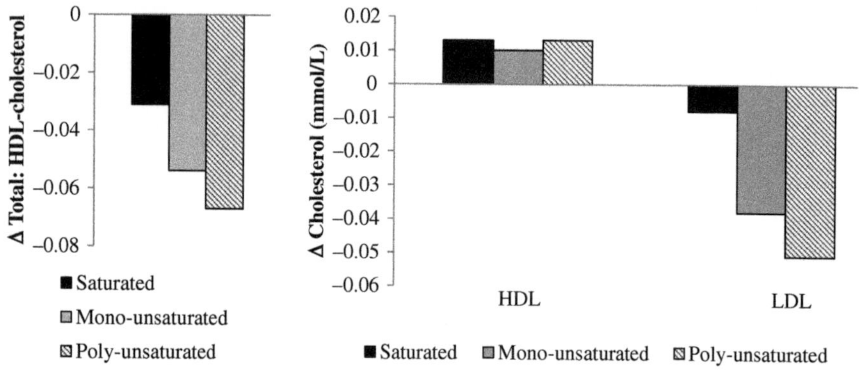

Figure 6.2 Predicted changes (Δ) in the ratio of serum total to HDL-cholesterol and in LDL- and HDL-cholesterol concentrations when *trans* fatty acids constituting 1% of energy are replaced iso-energetically with saturated, *cis*-monounsaturated and *cis*-polyunsaturated fatty acids.
Source: Mozaffarian D., Clarke R., 2009. Quantitative effects on cardiovascular risk factors and coronary heart disease risk of replacing partially hydrogenated vegetable oils with other fats and oils. Eur. J. Clin. Nutr. 63, S22–S33.

though TFAs are typically found in small amounts in foods that contain high amounts of SFAs, the relative effect of replacing TFA may be small when compared with replacing SFAs.

6.3.3.3 Dietary recommendations

The EFSA Panel considered evidence from national and international dietary reference intakes, such as those published by the US Institute of Medicine (2005). The Institute of Medicine (IoM) considered the evidence for SFAs, TFAs and blood cholesterol, and concluded that there was a positive linear relationship between these FAs and LDL-cholesterol concentrations. The IoM recommended that SFA and TFA intakes should be as low as possible. To reduce the intake of SFAs and TFAs and follow a 'healthful diet', this report suggested choosing lean cuts of meat, avoiding cakes and pastries, and eating smaller portions.

In addition to the IoM, the American Heart Association (AHA) published diet and lifestyle recommendations that included 'aiming for a desirable lipid profile' (Lichtenstein et al., 2006). With SFAs and TFAs identified as 'the strongest dietary determinants of elevated LDL-cholesterol', the AHA recommended intakes of SFAs and TFAs of <7 and <1% of energy respectively. As a result, the AHA identified animal fats and hydrogenated fats as primary sources of SFAs and TFAs.

In addition to the IoM and the AHA, the World Health Organization (WHO) issued a report in 2003, which the EFSA Panel considered as supporting evidence for the positive relationship between SFA and TFA intakes and cholesterol concentrations (WHO/FAO, 2003). Within this report, the WHO recommended that intakes of SFAs and TFAs should be limited to less than 10 and 1% of an individual's daily energy intake, respectively. Importantly, nutrient intake goals did not differentiate between *cis*- and *trans*-MUFA and PUFA. Similar to the IoM, the WHO derived its nutrient recommendations from RCTs, including a meta-analysis by Mensink and Katan (1992). In this meta-analysis of 27 RCTs, the effect of changes in the FA intake to changes in serum HDL-, LDL- and total cholesterol concentrations were evaluated. A replacement of 1% energy from carbohydrates with SFA increased LDL- and total cholesterol, whereas replacement with *cis*-PUFA decreased LDL-cholesterol. As a result, the WHO concluded that the most favourable strategy to lower CHD risk would be if SFAs and TFAs were replaced with UFA, with no decrease in total fat intake.

6.3.3.4 General discussion

Evidence for the health claim relating to MUFA and PUFA for reduced cholesterol and reduced risk of CHD originates from RCTs and observational studies, and has been considered in the context of dietary recommendations and various EFSA opinions for health claims. The totality of this evidence illustrates that there is a dose–response relationship between consumption of SFAs and TFAs, with an increase in LDL-cholesterol concentrations and a decrease in the total to HDL-cholesterol ratio. Given the link between cholesterol concentrations and cardiovascular disease (CVD), the detrimental role of SFAs and TFAs, compared with *cis*-MUFA and *cis*-PUFA, is considered of public health importance and thus worthy of communication to consumers by means of authorised health claims.

Although various sources of substantiating evidence are utilised (Table 6.1), the evidence for a relationship between FAs and blood cholesterol originates from RCTs, which have been aggregated in multiple meta-analyses to generate an overall estimate of risk. Based on the totality of the evidence, there is a general consensus that increasing amounts of mixtures of SFAs and TFAs in the diet will increase serum LDL-cholesterol concentrations. One potential mechanism behind this physiological response is a reduction in LDL receptor-mediated clearance, possibly by inhibition of gene expression (Mustad et al., 1997). Moreover, UFAs, *cis*-MUFA and *cis*-PUFA are preferential to SFA, *trans*-FA and carbohydrates due to their ability to increase LDL receptor-mediated clearance, increase reverse cholesterol transport (where cholesterol is transported from the peripheral tissue back to the liver for clearance) and act on important regulatory moieties (Fernandez and West, 2005). Although these meta-analyses are well-powered and frequently cited, these studies are now more than a decade old in some instances. A recent, well-powered systematic review and meta-analysis of observational studies ($n = 339{,}090$) addressing the intake of SFAs and TFAs and risk of all-cause mortality, CVD and type 2 diabetes concluded that although SFAs were not associated with all-cause mortality, CVD, CHD, ischemic stroke or type 2 diabetes, the evidence was limited by heterogeneity and methodological disparities (de Souza et al., 2015). In addition, industrially-produced TFAs, but not ruminant TFAs, were found to be associated with all-cause mortality, total CHD and CHD mortality. Importantly, this review was based on observational studies and so was unable to critically evaluate the effect of iso-energetic replacement of SFAs and TFAs with UFAs.

As a result, findings from Mensink et al. (2003) and Mozaffarian and Clarke (2009) are the most recent and relevant data to the present claim. Indeed, recent guidelines such as the European Guidelines on CVD prevention (Perk et al., 2012) still refer the meta-analysis by Mensink et al. (2003).

6.4 Other relevant legislation

In addition to the legislation surrounding the necessity to meet the conditions for use of the nutrition claim for 'High Unsaturated Fat', other legislation is indirectly relevant to the health claim in question. The implementation of nutrient profiles in the future—pursuant to Article 4 of Regulation (EC) No. 1924/2006 [EU Regulation (EC), 2006]—would be likely to introduce a restriction on the use of health claims on products containing SFA in excess of these profiles. However, the setting of nutrient profiles by the European Commission was postponed due to the complexity of the issue in relation to scientific challenges and potential economic impacts. Though nutrient profiles were one of the European Commission's key 23 initiatives for 2016, and included within a roadmap concerning partial reform of the Nutrition and Health Claims Regulation with a predicted completion date of June 2017, the European Parliament endorsed a call for abandoning nutrient profiles, and this will feed into the Commission discussions.

There are a number of parallels that can be drawn between the health claim in question and the Article 13.1 health claim that was given a positive EFSA opinion in 2011. This previous claim specifies that the replacement of SFAs with MUFAs and/or PUFAs is beneficial to cholesterol concentrations: 'Replacing saturated fats with unsaturated fats in the diet contributes to the maintenance of normal blood cholesterol levels' (EFSA Panel on Dietetic Products Nutrition and Allergies, 2011c) (MUFA and PUFA are unsaturated fats). Both claims are based on the same scientific substantiation and they both carry the same condition of 'High Unsaturated Fat'.

An additional claim for foods with a low or reduced content of SFAs with the wording 'Reducing consumption of saturated fat contributes to the maintenance of normal blood cholesterol levels' shares commonalties with the present health claim, yet is bound by different conditions (EFSA Panel on Dietetic Products Nutrition and Allergies, 2011b). Specifically, the claim must satisfy the conditions of use for the 'Low Saturated Fat' and 'Reduced' nutrition claims, as per the Annex of Regulation (EC) No. 1924/2006 [EU Regulation (EC), 2006]. To claim that a food is low in SFA, the sum of SFA and *trans*-FA in the product should not exceed 1.5 g/100 g for solids (or 0.75 g/100 mL for liquids) and the sum of SFA and *trans*-FA must not provide more than 10% of energy. Alternatively, if the conditions of the nutrition claim of reduced SFA are satisfied, then the reduction in content of SFA must be at least 30% compared with a similar product [EU Regulation (EC), 2006]. As a result, there are subtle differences in the implications for product development between these health claims.

6.5 Consumer issues

EU regulations on health claims are in place to protect the consumer from misleading and false claims and are designed to facilitate consumers to make an informed and health-conscious purchasing choice. To ensure that health claims meet these aims there are two consumer-related issues that should be discussed. First, the ability of the consumer to understand the health claim should be considered and second, the benefits of the health claim to the consumer should be objectively evaluated.

6.5.1 Understanding the health claim

The wording of the present claim was proposed by the applicant as 'Replacing a fat rich in saturated/*trans* fatty acids by a fat rich in unsaturated fatty acids helps to reduce LDL cholesterol. LDL cholesterol is a cardiovascular risk factor'. The proposed wording was modified by EFSA to facilitate consumer understanding and reflect the scientific evidence: 'Consumption of saturated fat increases blood cholesterol concentrations; consumption of mono- and/or polyunsaturated fat in replacement of saturated fat has been shown to lower/reduce blood cholesterol. Blood cholesterol lowering may reduce the risk of (coronary) heart disease'. It was then further modified during the authorisation procedures following discussion with Member States and further discussion with EFSA, to the final wording: 'Replacing saturated fats with unsaturated fats in the diet has been shown to lower/reduce blood cholesterol. High cholesterol is a risk factor in the development of coronary heart disease' (EFSA Panel on Dietetic Products Nutrition and Allergies, 2011a).

As specified by EU Regulation 1924/2006 Recital 15, 'the use of nutrition and health claims should only be permitted if the average consumer can be expected to understand the beneficial effects as expressed in the claim'. Within this context the average consumer is defined as someone 'who is a reasonably well informed and reasonably observant and circumspect, taking into account social, cultural and linguistic factors'. Although the research in this space is limited, a consistent determinant of consumer understanding is education level, with well-educated individuals being more likely to understand the association between diet and disease (Cotunga et al., 1992) and less-well educated consumers being more prone to inappropriate interpretations of health messages. According to a study conducted by the US Food and Drug Administration, the presence of a front-of pack health claim can induce phenomena such as the 'halo effect'—where the product is rated higher on additional attributes that are not mentioned in the claim and also the 'magic bullet' effect—where inaccurate health benefits are associated with the product (Roe et al., 1999).

Two reviews in this area concluded that health claims are perceived by consumers as difficult to understand (Williams, 2005; Wills et al., 2012). In short, consumers prefer simple and easy-to-understand information on the front-of pack, with more detailed information provided on back-of pack (Williams, 2005). Wills et al. (2012) suggested consumer understanding varied according to the nature of the claim, knowledge about the claim or the substance in the claim, familiarity with the product and the claim and terminology used and the respondent's country of origin. The challenge of the European Commission is how to accurately capture the scientific meaning into language that is easily understood by the consumer. To facilitate this process, co-operation between the scientific community, regulatory bodies and the food industry is needed.

6.5.2 What are the benefits to consumers?

The benefits associated with the use of a health claim are centred on the ability of the consumer to make an informed choice that has the potential to improve their health. Regarding the present claim, products bearing the claim are low in SFA and TFA and high in UFA which is considered beneficial to the reduction of blood LDL-cholesterol concentrations. Nonetheless, this benefit should be contextualised within the framework of the diet as a whole, as without adherence to a healthy-balanced diet, any beneficial effect of this product is likely to be minimal.

From the food industry's perspective, health claims are designed to improve the marketing impact and subsequent sales of the product bearing the claim. Retail trends suggest that the food industry often provides an added benefit to the consumer above that of the health claim to drive sales. Such examples include the product being organic or fair trade. To protect the consumer from false advertising, European Commission Regulations provide the benefit to the consumer of being confident in the scientific basis for the health claim.

6.6 Potential impact of the claim on product development

Authorisation of health claims could have a substantial impact on product development for the food product to meet the conditions of use. For example, the recent authorisation of a health claim pertaining to foods with a low or reduced content of SFAs aligns very closely with a widespread movement by the food industry to reduce the SFA content of their products. For example, United Biscuits have reduced the SFA content of Hula Hoops and Skips by 80% since 2005 (Food and Drink Federation, 2016). In addition, the UK Department of Health (2013) launched the Responsibility Deal Saturated Fat Reduction Pledge in 2013 which aimed to remove thousands of tons of saturated fat from the UK nation's diet. With 13 companies involved, representing almost half of the food manufacturing and retail industry, this pledge aims to have significant implications for public health. In addition, health claims could encourage product innovation through the development of functional foods which is particularly relevant for the use of certain vitamins and minerals (Richardson and Eggersdorfer, 2015). In the present case, the claim pertaining to low SFA and TFA can only be used in fats and oils which makes the most directly applicable food product.

Reformulation would not be necessary for a product to bear the present claim if it already complied with the conditions of the 'High Unsaturated Fat' nutrition claim, that is 'at least 70% of the FAs present in the product derive from unsaturated fat under the condition that unsaturated fat provides more than 20% of energy of the product'. [EU Regulation (EC), 2010]. Some readily available spreads may already meet this condition given that spreads undergo heavy processing in their production and can be blended to have as little as 5% total fat by weight. Taking the UKs Flora 'Proactive Olive spread' as an example, 77% of the FAs are UFA and these UFAs provide 76% of the total energy. In contrast, if the fat or oil in question did not meet the conditions of the claim then reformulation would be necessary. In the case of fats and oils, an increase in UFA would have textural and functional implications including a softening of the product and potential difficulties when used for frying and baking.

6.7 Future trends

The global burden of obesity and chronic disease is increasing. As a result, the mounting pressure to improve modifiable risk factors such as diet will help drive improvement in the nutrient profiles of food products. Although food manufacturers are utilising reformulation strategies to achieve these changes, there is a growing consensus in both the scientific and health professionals' community that food constituents such as FA, are not a representative basis for recommendations given that food choices relate to whole foods rather than to nutrients. As such, future trends may focus on whole foods and dietary patterns, as outlined in the EFSA scientific opinion on establishing Food-Based Dietary Guidelines (EFSA Panel on Dietetic Products Nutrition and Allergies, 2010). However, the present EU labelling system is based on nutrients and hence a nutrient-centric approach will remain integral to health claim regulations for the foreseeable future.

The EFSA Panel based the health claim in question on substantiating evidence for LDL-cholesterol concentrations. Due to the transitory nature of blood cholesterol, the use of this marker in predicting CVD is more limited than some of the more novel, holistic markers of CVD such as arterial stiffness (Vlachopoulos et al., 2010). Current research linking SFA, *cis*-MUFA and *cis*-PUFA and arterial stiffness is limited (Livingstone et al., 2013); however, with the addition of future studies there is potential to submit further claims to EFSA in relation to these markers.

6.8 Conclusions

The evidence in support of the relationship between 'Replacing saturated fats with unsaturated fats in the diet' and reduction of LDL-cholesterol concentrations is strong. However, product reformulation may be required for some types of spreads to meet the conditions of use. With regard to consumer understanding, future research is warranted to optimise the clarity of heath claims while still maintaining their scientific basis.

6.9 Sources of further information and advice

Following are the three suggested sources of further information regarding FAs in foods, heart disease and EU health claim legislation.

6.9.1 Saturated fat in foods

A publication by Tabot (2011), entitled 'Reducing saturated fats in food's provides an informative overview of SFA including sources and health aspects, and an in-depth strategic evaluation of how to reduce the SFA content of various foods: milk and dairy products, butchered and processed meat products, savoury snacks and fried foods, biscuits, pastry, chocolate, ice-cream and sauces.

6.9.2 British Heart Foundation

The British Heart Foundation is a UK charity that provides advice to the public on heart disease and funds scientific research to better understand and treat the condition. Its website (https://www.bhf.org.uk/) includes information on heart health including risk factors such as raised LDL-cholesterol, scientific blogs from leading researchers in the field of heart health and information on their current research projects.

6.9.3 European Food Safety Authority

ESFA's website (http://www.efsa.europa.eu/), publishes information pertinent to European regulations on health claims including current news and events. Opinions from the Scientific committee/panel on health claims from Article 13, 13.5, 14 and 19 are also published online. For further information on the requirements for submission of a claim to EFSA see the following report: 'Scientific and technical guidance for the preparation and presentation of an application for authorisation of a health claim (revision 1)'.

References

Cotunga, N., Subar, A.F., Heimendinger, J., Kahle, L., 1992. Nutrition and cancer prevention knowledge, beliefs, attitudes, and practices: the 1987 National Health Interview Survey. J. Am. Diet. Assoc. 92, 963–968.

de Souza, R.J., Mente, A., Maroleanu, A., Cozma, A.I., Ha, V., Kishibe, T., Uleryk, E., Budylowski, P., Schünemann, H., Beyene, J., Anand, S.S., 2015. Intake of saturated and *trans* unsaturated fatty acids and risk of all cause mortality, cardiovascular disease, and type 2 diabetes: systematic review and meta-analysis of observational studies. Br. Med J. h3978, 351.

Department of Health, 2013. Thousands of tonnes of saturated fat to be taken out of the nation's diet. Available from: https://www.gov.uk/government/news/thousands-of-tonnes-of-saturated-fat-to-be-taken-out-of-the-nations-diet.

EFSA, 2015. Claims on disease risk reduction and child development or health under Article 14. Available from: http://www.efsa.europa.eu/en/topics/topic/article14.

EFSA Panel on Dietetic Products Nutrition and Allergies, 2004. EFSA Opinion of the Scientific Panel on Dietetic Products, Nutrition and Allergies on a request from the Commission related to the presence of *trans* fatty acids in foods and the effect on human health of the consumption of *trans* fatty acids. EFSA J. 81, 1–49.

EFSA, 2010. Scientific opinion on dietary reference values for fats, including saturated fatty acids, polyunsaturated fatty acid monounsaturated fatty acids, trans fatty acids, and cholesterol. EFSA J. 8 (3), 1461.

EFSA Panel on Dietetic Products Nutrition and Allergies, 2011a. Scientific Opinion on the substantiation of a health claim related to "low fat and low *trans* spreadable fat rich in unsaturated and omega-3 fatty acids" and reduction of LDL-cholesterol concentrations pursuant to Article 14 of Regulation (EC) No. 1924/2006. EFSA J. 9, 2168.

EFSA Panel on Dietetic Products Nutrition and Allergies, 2011b. Scientific Opinion on the substantiation of health claims related to foods with reduced amounts of saturated fatty acids (SFAs) and maintenance of normal blood LDL-cholesterol concentrations (ID 620, 671, 4332) pursuant to Article 13(1) of Regulation (EC) No. 1924/2006. EFSA J. 9, 2062.

EFSA Panel on Dietetic Products Nutrition and Allergies, 2011c. Scientific Opinion on the substantiation of health claims related to the replacement of mixtures of saturated fatty acids (SFAs) as present in foods or diets with mixtures of monounsaturated fatty acids (MUFAs) and/or mixtures of polyunsaturated fatty acids (PUFAs), and maintenance of normal blood LDL-cholesterol concentrations (ID 621, 1190, 1203, 2906, 2910, 3065) pursuant to Article 13(1) of Regulation (EC) No. 1924/2006. EFSA J. 9, 2069.

EU Regulation (EC), 2006. Commission Regulation (EU) No. 1924/2006 of the European Parliament and of the council of 20 December 2006 on nutrition and health claims made on foods by the European Food Safety Authority. Off. J. Eur. Union 404, 9–25.

EU Regulation (EC), 2010. Commission Regulation (EU) No. 116/2010 of 9 February 2010 amending Regulation (EC) No. 1924/2006 of the European Parliament and of the Council with regard to the list of nutrition claim. Off. J. Eur. Union 37, 16–18.

EU Regulation (EC), 2014. Commission Regulation (EU) No. 1226/2014 of 17 November 2014 on the authorisation of a health claim made on foods and referring to the reduction of disease risk. Off. J. Eur. Union 331, 3–5.

European Commission, 2007. Guidance on the implementation of Regulation No. 1924/2006 on nutrition and health claims made on foods, conclusions of the standing committee on the food chain and animal health Available from: http://ec.europa.eu/food/safety/docs/labelling_nutrition_claim_reg-2006-124_guidance_en.pdf.

Fernandez, M.L., West, K.L., 2005. Mechanisms by which dietary fatty acids modulate plasma lipids. J. Nutr. 135, 2075–2078.

Food and Drink Federation, 2016. Industry case studies. Available from: http://www.fdf.org.uk/casestudies.aspx.

Institute of Medicine, 2005. Dietary Reference Intakes for Energy, Carbohydrate, Fibre, Fat, Fatty Acids, Cholesterol, Protein, and Amino Acids (Micronutrients). The National Academies Press, Washington, DC.

Lichtenstein, A.H., Appel, L.J., Brands, M., Carnethon, M., Daniels, S., Franch, H.A., Franklin, B., Kris-Etherton, P., Harris, W.S., Howard, B., Karanja, N., Lefevre, M., Rudel, L., Sacks, F., Van Horn, L., Winston, M., Wylie-Rosett, J., 2006. Diet and Lifestyle Recommendations Revision 2006: a scientific statement from the American Heart Association Nutrition Committee. Circulation 114, 82–96.

Lichtenstein, A.H., Ausman, L.M., Jalbert, S.M., Schaefer, E.J., 1999. Effects of different forms of dietary hydrogenated fats on serum lipoprotein cholesterol levels. N. Engl. J. Med. 340 (25), 1933–1940.

Livingstone, K.M., Givens, D.I., Cockcroft, J.R., Pickering, J.E., Lovegrove, J.A., 2013. Is fatty acid intake a predictor of arterial stiffness and blood pressure in men? Evidence from the Caerphilly Prospective Study. Nutr. Metab. Cardiovasc. Dis. 23, 1–7.

Mensink, R.P., Katan, M.B., 1992. Effect of dietary fatty acids on serum lipids and lipoproteins. A meta-analysis of 27 trials. Arterioscler. Thromb. Vasc. Biol. 12, 911–919.

Mensink, R.P., Zock, P.L., Kester, A.D., Katan, M.B., 2003. Effects of dietary fatty acids and carbohydrates on the ratio of serum total to HDL cholesterol and on serum lipids and apolipoproteins: a meta-analysis of 60 controlled trials. Am. J. Clin. Nutr. 77, 1146–1155.

Mozaffarian, D., Clarke, R., 2009. Quantitative effects on cardiovascular risk factors and coronary heart disease risk of replacing partially hydrogenated vegetable oils with other fats and oils. Eur. J. Clin. Nutr. 63, S22–S33.

Mustad, V.A., Etherton, T.D., Cooper, A.D., Mastro, A.M., Pearson, T.A., Jonnalagadda, S.S., Kris-Etherton, P.M., 1997. Reducing saturated fat intake is associated with increased levels of LDL receptors on mononuclear cells in healthy men and women. J. Lipid Res. 38, 459–468.

Perk, J., De Backer, G., Gohlke, H., Graham, I., Reiner, Ž., Verschuren, M., Albus, C., Benlian, P., Boysen, G., Cifkova, R., Deaton, C., Ebrahim, S., Fisher, M., Germano, G., Hobbs, R., Hoes, A., Karadeniz, S., Mezzani, A., Prescott, E., Ryden, L., Scherer, M., Syvänne, M., Scholte Op Reimer, W.J.M., Vrints, C., Wood, D., Zamorano, J.L., Zannad, F., 2012. European guidelines on cardiovascular disease prevention in clinical practice (version 2012): the fifth joint task force of the European Society of Cardiology and other societies on cardiovascular disease prevention in clinical practice (constituted by representatives of nine societies and by invited experts) developed with the special contribution of the European Association for Cardiovascular Prevention & Rehabilitation (EACPR). Eur. Heart J. 33, 1635–1701.

Richardson, D.P., Eggersdorfer, M., 2015. Opportunities for product innovation using authorised European Union health claims. Int. J. Food Sci. Technol. 50, 3–12.

Roe, B., Levy, A.S., Derby, B.M., 1999. The impact of health claims on consumer search and product evaluation outcomes: results from FDA experimental data. J. Public Policy Mark. 18, 89–105.

Sadler, M. (Ed.), 2015. Foods, Nutrients and Food Ingredients with Authorised EU Health Claims. Woodhead Publishing, Cambridge.

Tabot, G., 2011. Reducing Saturated Fats in Foods. Woodhead Publishing, Cambridge.

Vlachopoulos, C., Aznaouridis, K., Stefanadis, C., 2010. Prediction of cardiovascular events and all-cause mortality with arterial stiffness: a systematic review and meta-analysis. J. Am. Coll. Cardiol. 55, 1318–1327.

WHO/FAO, 2003. Expert report: diet, nutrition and prevention of chronic diseases. Report of a Joint WHO/FAO Expert Consultation. WHO Technical Report Series. WHO Technical Report Series 916.

Williams, P., 2005. Consumer understanding and use of health claims for foods. Nutr. Rev. 63, 256–264.

Wills, J.M., Storcksdieck genannt Bonsmann, S., Kolka, M., Grunert, K.G., 2012. European consumers and health claims: attitudes, understanding and purchasing behaviour. Proc. Nutr. Soc. 71, 229–236.

Part Three

Health Claims Based on Emerging Science and Proprietary Claims

Authorised EU health claim for reformulated, non-alcoholic, acidic drinks

7

Michele J. Sadler
Rank Nutrition Ltd, Bethersden, Kent, United Kingdom

7.1 Introduction

Oral diseases are a major public health problem, even though their aetiology is well known. Maintenance of dental health is an important consideration for all age groups. Teeth are important for continued consumption of a varied diet selected from a range of food groups, through the lifespan and for the enjoyment of food (Moynihan and Petersen, 2004). Two key dental diseases are dental caries and dental erosion.

Diet is a key component in the aetiology of dental caries, as it provides fermentable substrate for bacteria that resides in dental plaque. Dental caries ultimately arises from acid produced by these bacteria following fermentation of dietary sugars or of sugars released from fermentable carbohydrates in the mouth by the action of amylase. If plaque pH falls below 5.5, the enamel becomes more soluble increasing the potential for demineralisation of the tooth enamel. Over time this can lead to a carious lesion or cavity which is a late stage of the disease. However, a beneficial factor is saliva which is stimulated by chewing and by organoleptic attributes of foods and drinks. Saliva has a clearing action in the mouth. As it is supersaturated with calcium and phosphate, saliva has a neutral pH of 7, and thus has a buffering effect on the teeth which helps to protect them from demineralisation. Saliva also promotes remineralisation of the enamel, by carrying calcium and phosphate to the tooth surfaces. The balance of demineralisation (tooth lesions) and remineralisation (tooth repair) determines the clinical outcome over time, and whether dental caries develops or not.

In contrast, dental erosion is the irreversible loss of dental enamel that is caused by acid from food and drink per se, rather than from activity of the plaque bacteria. The diet provides extrinsic acids, and any food or drink with a pH below 5.5 may contribute to dental erosion.

This chapter considers an authorised EU health claim relating to the maintenance of tooth mineralisation which is relevant to the processes that lead to both dental caries and dental erosion.

Foods, Nutrients and Food Ingredients with Authorised EU Health Claims. http://dx.doi.org/10.1016/B978-0-08-100922-2.00007-3
Copyright © 2018 Elsevier Ltd. All rights reserved.

7.2 Health claim application

The application, by GlaxoSmithKline Services, was for an Article 13.5 claim which included a request for the protection of proprietary data. The proposed claim was for a range of modified, non-alcoholic, fruit juice drinks, branded as Toothkind, a variant of Ribena drinks. The health relationship was the maintenance of healthy teeth when the drinks are used as an alternative to soft drinks or sugar-sweetened beverages, and fruit juices, in the context of a balanced diet.

To fulfil the proposition of the Toothkind drinks, it was considered that the drinks should have minimal erosive and cariogenic potential. This was achieved by controlling the calcium to acid molar ratio, the pH and the content of fermentable carbohydrate in the drinks.

The applicant's proposed wording for the claim was 'Toothkind drinks help to maintain healthy teeth'. However, EFSA considered the beneficial physiological effect to be a reduction in tooth demineralisation which was established after clarification with the applicant. The evidence was assessed against this effect, although in the context of EFSA's proposed claim wording and the type of claim, the Commission later sought further clarification on this issue (Section 7.6).

7.3 Characterisation

Toothkind drinks consist of fruit juice which supplies all of the carbohydrate and most of the acidity, to which calcium is added. They are characterised as having a:

- Molar ratio of calcium to acid between 0.3 and 0.8. The actual amount of calcium added is determined by the measured titratable acidity of the product. A lower content of added calcium would reduce efficacy while a higher content might result in calcium precipitation.
- A pH in the range of 3.7–4.0. The lower limit is determined on the basis of efficacy and the upper limit on the basis of microbiological and taste effects.
- A fermentable carbohydrate content of less than 1 g/100 mL (range 0–0.9 g/100 mL).

7.3.1 Characterisation of comparator products

The evidence used to substantiate the claim was based on comparisons of Toothkind drinks with soft drinks and fruit juices. Whereas fruit juices mostly contain organic acids, other types of acids may be added to soft drinks. Thus it was also necessary to characterise the comparator products. The two types of comparator products had:

- A calcium to acid molar ratio of 0.015–0.1.
- A pH in the range of 2.6–3.6.
- A carbohydrate content of 8–12 g/100 mL (mostly sugars).

7.4 Target group

The target group for the claim is the general population including both adults and children.

7.5 Evidence substantiating the claim

In accordance with EFSA guidance for health claim applications, the applicant conducted a systematic review of the literature. The literature searches resulted in a total of 35 pertinent studies, of which 10 were published human intervention studies, 15 were unpublished human intervention studies and 10 were non-human studies.

There are a number of techniques available to assess the potential of foods and drinks to demineralise the teeth. One such method is the in situ enamel slab technique in which subjects wear a removable appliance that is fitted with slabs of human or synthetic enamel. The overall impact of a food or drink can be tested on demineralisation potential, taking account of protective as well as detrimental factors. Three unpublished studies were identified that used this technique, the endpoint being the micro-hardness of the enamel. The protocols tested ex vivo exposure to beverages for periods of 2 min, 7 times a day for up to 21 days. Lower frequencies of exposure were not investigated. Toothkind drinks resulted in significantly less softening of the enamel compared with a sucrose solution (10 g/100 mL), more softening than in comparison with mineral water and equivalent softening compared with sorbitol (10 g/100 mL) (EFSA, 2010). Sorbitol is a non-cariogenic sugar alcohol as it is not metabolised by plaque bacteria, that is thus used as a negative control. The extent of enamel softening increased throughout the studies for both the Toothkind drinks and the comparator solutions.

A further approach to determine the effects of foods and drinks on dental health is to measure their effects on proxy markers of dental caries and dental erosion. Acid production resulting from carbohydrate fermentation by the plaque bacteria reduces the pH of plaque, and this endpoint is an indirect measure of demineralisation. It can be measured by a variety of techniques and different endpoints can be used. A food is considered to be harmful to teeth if plaque pH falls below 5.7–5.5 within 30 min after consumption (determined by measuring the area under the Stephan curve). Ten unpublished studies and one published study used this approach and the main measures were mean plaque pH 10 min after exposure to the drinks and mean minimum plaque pH (EFSA, 2010). In 10 of the studies Toothkind drinks showed significantly smaller reductions in plaque pH compared with a sucrose solution (10 g/100 mL) and with a traditional juice drink (4.5 g sugars/100 mL) (Toumba and Duggal, 1999), though not with a conventional juice drink containing less than 1 g sugars/100 mL. Crucially, mean minimum plaque pH and mean minimum plaque pH 10 min after exposure did not fall below pH 5.7, the critical threshold of pH below which demineralisation starts to occur. In contrast, the mean minimum plaque pH following sucrose solution was 5.1–5.8. EFSA considered that the studies indicated that the sucrose solution had the potential to enable demineralisation, though the studies did not measure actual net demineralisation, only the proxy marker of plaque pH.

The potential of Toothkind drinks to protect against erosion was measured in 10 randomised, controlled cross-over trials. Dental enamel slabs were worn in the mouth throughout the test periods and the endpoint was enamel loss. In most of the studies the subjects consumed 4 × 250 mL servings of the drinks over the course of a day, for periods ranging from 1 to 15 days. Frequencies of exposure were not investigated.

In nine of the studies (two unpublished) Toothkind juice drinks showed significantly less erosive potential than traditional juice drinks (Finke et al., 2001; Hughes et al., 1999a,b, 2002; West et al., 1999, 2003, 2004) and either similar or slightly higher erosive potential than water. In one study, Toothkind carbonated drinks showed no significant difference to traditional soft drinks (8–12 g sugars/100 mL) (Hunter et al., 2003).

A study that measured change in pH at the tooth surface over a period of 15 min was considered inadmissible evidence by EFSA as this was not an appropriate method to assess the erosive potential.

A series of in vitro studies provided further supportive evidence that Toothkind juice drinks and Toothkind carbonated drinks showed lower erosive potential than traditional juice or soft drinks (EFSA, 2010). While this supported the evidence from the in situ studies for a reduced erosive potential of Toothkind juice drinks against traditional juice drinks, the in situ studies had not shown a benefit of Toothkind carbonated drinks against traditional soft, carbonated drinks (8–12 g sugars/100 mL).

From the overall evidence base, EFSA concluded that Toothkind drinks reduced tooth demineralisation compared with a sugar solution (10 g/100 mL) for a frequency of 7 times per day, with no evidence investigating exposure at lower frequency. EFSA also concluded that Toothkind drinks reduced tooth demineralisation compared with traditional juice drinks at an exposure of 4 times per day, providing evidence for a reduced erosive potential.

The NDA panel thus concluded that a cause and effect relationship was established for traditional juice drinks consumed 4 times per day and tooth demineralisation due to its erosive potential, and that replacing them with Toothkind juice drinks may contribute to a reduced demineralisation. They also concluded that a cause and effect relationship was established for the consumption of traditional soft drinks (8–12 g sugars/100 mL) when consumed 7 times per day and tooth demineralisation through reduced plaque pH, and that replacement with Toothkind juice drinks may contribute to reduced demineralisation (EFSA, 2010).

Though the applicant had proposed conditions of use for substitution of a serving of a traditional soft drink with a Toothkind drink, the effect of individual servings had not been tested, and EFSA therefore concluded in the initial opinion that the evidence only supported the proposed conditions of use in frequent consumers of traditional soft drinks and juice drinks (EFSA, 2010).

The EFSA opinion also concluded that the scientific assessment of the claim was performed as a comparison between the drinks for which the claim is intended and traditional juice drinks and traditional soft drinks (EFSA 2010).

The wording for the claim proposed by EFSA was ‘Frequent consumption of typical juice drinks and sugar-containing acidic non-alcoholic beverages may contribute to tooth demineralisation; consumption of Toothkind juice drinks in replacement of typical juice drinks and sugar-containing acidic, non-alcoholic beverages may help to reduce tooth demineralisation’. However, following discussion during the authorisation procedures, and further advice from EFSA, the final authorised wording was different to this (Section 7.6).

7.6 Authorisation of the claim

The EFSA opinion was published in 2010, but the claim was not fully authorised until September 2013 (EU, 2013). In the intervening period a further statement was published by EFSA addressing three questions that arose during the authorisation procedures from discussions between the EU Commission and Member States representatives (EFSA, 2011).

Firstly, EFSA was asked whether the evidence measuring the potential for tooth demineralisation by measuring pH changes, for example rather than measuring tooth demineralisation directly, was suitable to demonstrate the beneficial effect, that is reduction of tooth demineralisation.

On this point EFSA was clear that the potential for demineralisation of enamel that is due to the fermentation of sugars in beverages causing a reduction in dental plaque pH, can be determined following a single consumption of a beverage and acute measurement of plaque pH. However, multiple intakes of the test beverages are required to demonstrate the effect, because a single daily consumption would not cause net demineralisation due to the capacity for repair of the enamel and the dynamic process between demineralisation and remineralisation. This type of demineralisation can lead to dental caries, and EFSA confirmed its conclusion that Toothkind drinks have little or no potential for demineralisation because plaque pH is not reduced below 5.7. In contrast, demineralisation of enamel that is due to the acidic nature of the beverage itself results in tooth erosion. The evidence to support both of these effects was provided by in situ studies in which subjects wore dental enamel samples over periods of days or weeks and the test and control products were consumed at a frequency of 7 doses per day.

Secondly, EFSA was asked whether the beneficial effect would be expected to occur with less frequent consumption, for example with replacement of one serving of typical juice drinks or typical sugar-containing soft drinks with a serving of Toothkind drinks. On this point EFSA replied that a beneficial effect on maintaining tooth mineralisation would be expected with the substitution of one or more soft drinks or typical juice drinks with equivalent servings of Toothkind drinks, particularly in people who frequently consume sources of sugar and acids from other foods and beverages that can contribute to tooth demineralisation.

The third issue related to the proposed wording of the claim. The application was for an Article 13.5 claim which is essentially a ‘general function claim’. However, because tooth demineralisation is a risk factor in the development of dental caries EFSA’s proposed wording was more akin to an Article 14.1(a) disease risk reduction claim. The Commission wanted to examine alternative wordings that would not lead to confusion on the part of consumers regarding the type of health claim. EFSA’s response was that ‘maintenance of tooth mineralisation’ has a similar meaning to ‘reduction of tooth demineralisation’.

During the authorisation processes, careful consideration was given to consumer understanding of the claim wording, particularly over terms such as ‘reformulated’ and ‘acidic’. The conditions of use were also debated in terms of whether or not to refer to ‘frequent consumers’. A concern was also raised about disadvantaging mineral waters.

In the final analysis, the officially authorised wording of the claim was agreed as:

Replacing sugar-containing, acidic drinks, such as soft drinks (typically 8-12 g sugars/100 ml), with reformulated drinks contributes to the maintenance of tooth mineralisation.

This is the wording in the Regulation authorising the claim (EU, 2013) and that is indicated on the Community List of Claims (Commission Website, 2017).

The claim was specified for use on:

Reformulated, non-alcoholic, acidic drinks with:

- <1 g fermentable carbohydrate (sugars and other carbohydrates except polyols) /100 mL.
- Calcium in the range 0.3–0.8 mol/mol acidulant.
- pH between 3.7 and 4.0.

The conditions of use were:

In order to bear the claim, reformulated acidic drinks shall comply with the description of the food subject to the claim.

No warning labels were required and no safety issues were identified.

7.6.1 Proprietary nature of the claim

EFSA concluded that it could not have reached its determination that a cause and effect relationship was established without taking account of the unpublished, proprietary studies submitted by the applicant. The claim was therefore granted protection of proprietary data for 5 years, expiring on 23 September 2018. Due to company changes, the applicant named in the Community list is 'Lucozade Ribena Suntory Limited and its affiliates'.

7.7 Consumer issues

A key requirement of the Nutrition and Health Claims Regulation (EC, 2006) is that a significant quantity of the product that will produce the nutritional or physiological effect can reasonably be expected to be consumed [Article 5.1(d)]. As part of its assessment, EFSA also considers whether such a quantity can reasonably be consumed as part of a balanced diet.

It is clear from EFSA's additional statement (EFSA, 2011) that substitution of one or more soft drinks or typical juice drinks with equivalent servings of Toothkind drinks would be expected to be beneficial, particularly for those who consume other sources of sugar and acids over the course of a day, and that each replacement is likely to be beneficial.

7.8 Use of the claim

Though the claim has been authorised for use in the European Union, it is not currently in use as the Toothkind variant of Ribena-branded drinks is no longer on sale.

As is often the case with new product launches, the concept of the Toothkind formula was received positively by consumers, but sales over the years did not prove as popular as the core Ribena drink. However, in regions where the Toothkind variant was more popular, for example in Ireland, it continued to be sold until 2014.

As consumer trends have moved towards lower sugar variants, the Ribena brand has focused its marketing and drink formulations on no added sugar variants such that the Toothkind formulation is no longer produced.

7.9 Related EFSA opinion

In 2011 EFSA issued a scientific opinion on a claim application submitted via the Netherlands for an Article 14.1(a) disease risk reduction claim (EFSA, 2011a). This was for acidic calcium-containing fruit juices, in place of fruit juices, and reduction of tooth demineralisation as a risk factor for dental erosion. EFSA considered that the food, that is acidic calcium-containing fruit juices and comparative products were sufficiently characterised. EFSA also considered that reducing demineralisation in relation to the erosive potential of a food or drink is a beneficial physiological effect provided that it is not accompanied by tooth demineralisation arising from fermentation of carbohydrates and thus the production of acid in dental plaque.

The applicant submitted 37 studies to substantiate the claim. This included 15 observational studies, 3 reviews, 6 intervention studies, 12 in vitro studies and 1 animal study. However, none of the studies addressed the potential of fruit juices, with or without added calcium, for demineralisation of dental enamel resulting from acid production in plaque following fermentation of sugars. EFSA therefore concluded that the possible effects of both acids and sugars had not been properly assessed, such that a cause and effect relationship could not be established. A negative opinion was therefore issued (EFSA, 2011a).

7.10 Future trends

Diet has an important role in maintaining dental health, and there is a general need for more research to understand the impact of different foods and drinks on dental endpoints, regarding both appropriate amounts and frequency of consumption. For example a recent systematic review investigated the effect of dried fruit on dental health. Despite the generally poor evidence available, some countries give advice to restrict the timing of consumption of dried fruit, for example limiting consumption to mealtimes only. However, it was found that there is currently unconvincing and limited evidence to draw conclusions about the impact of dried fruit on dental health and that it is therefore premature to formulate evidence-based advice on the unsuitability of dried fruits as a snack. The review concluded that there is a need for further, high quality research, to improve our understanding of the role of specific foods and drinks in relation to dental health and to underpin robust, evidence-based advice. A number of suggestions for further research requirements are made in the review (Sadler, 2016).

7.11 Conclusions

A claim for the maintenance of tooth mineralisation has been authorised for use in the European Union on a range of reformulated, non-alcoholic, acidic drinks containing less than 1 g fermentable carbohydrate/100 mL (sugars and other carbohydrates except polyols), 0.3–0.8 mol calcium/mol acidulant, and with a pH in the range 3.7–4.0 (Toothkind drinks). The claim has proprietary protection for a period of 5 years such that it can only be used by the applicant. This expires in September 2018. However, due to market changes and differing consumer demands, the Toothkind formulation is no longer produced and the claim is not in use.

The maintenance of tooth mineralisation is important for dental health, in relation to both dental caries and dental erosion. To substantiate the claim for this innovative product, a large body of research evidence was compiled with the product. To improve our understanding of the role of specific foods and drinks in relation to dental health more of this type of research is warranted.

References

Commission Website, 2017. Available from: http://ec.europa.eu/food/safety/labelling_nutrition/claims/register/public/?event=register.home.

EFSA, 2010. Scientific Opinion on the substantiation of a health claim related to toothkind drinks and reduction of tooth demineralisation pursuant to Article 13(5) of Regulation (EC) No. 1924/2006. EFSA Panel on Dietetic Products, Nutrition and Allergies (NDA). EFSA J. 8 (12), 1884.

EFSA, 2011. Scientific Opinion—Statement on 'Toothkind' juice drinks. EFSA Panel on Dietetic Products, Nutrition and Allergies (NDA). EFSA J. 9 (7), 2293.

EFSA, 2011a. Scientific Opinion on the substantiation of a health claim related to calcium-containing fruit juices and the reduction of tooth demineralisation pursuant to Article 13(5) of Regulation (EC) No. 1924/2006. EFSA Panel on Dietetic Products, Nutrition and Allergies (NDA). EFSA J. 9 (2), 19.

EC, 2006. Regulation (EC) No. 1924/2006 of the European Parliament and the Council of 20 December 2006 on Nutrition and Health Claims made on foods, as amended by Corrigenda published in Official Journal of the European Union 18.1.2007, L12/3.

EU, 2013. Commission Regulation (EU) No. 851/2013 of 03-09-2013 authorising certain health claims made on foods, other than those referring to the reduction of disease risk and to children's development and health, amending Regulation (EU) No. 432/2012.

Finke, M., Hughes, J.A., Parker, D.M., Jandt, K.D., 2001. Mechanical properties of in situ demineralised human enamel measured by AFM nanoindentation. Surf. Sci. 491, 456–467.
Hunter, M.L., Hughes, J.A., Parker, D.M., West, N.X., Newcombe, R.G., Addy, M., 2003. Development of low erosive carbonated fruit drinks. 1. Evaluation of two experimental orange drinks in vitro and in situ. J. Dent. 31, 253–260, UK.
Hughes, J.A., West, N.X., Parker, D.M., Newcombe, R.G., Addy, M., 1999a. Development and evaluation of a low erosive blackcurrant juice drink in vitro and in situ. 1. Comparison with orange juice. J. Dent. 27, 285–289.
Hughes, J.A., West, N.X., Parker, D.M., Newcombe, R.G., Addy, M., 1999b. Development and evaluation of a low erosive blackcurrant juice drink 3. Final drink and concentrate, formulae comparisons in situ and overview of the concept. J. Dent. 27, 345–350.
Hughes, J.A., Jandt, K.D., Baker, N., Parker, D., Newcombe, R.G., Eisenburger, M., Addy, M., 2002. Further modification of soft drinks to minimise erosion. Caries Res. 36, 70–74.
Moynihan, P., Petersen, P.E., 2004. Diet, nutrition and the prevention of dental diseases. Public Health Nutr. 7 (1A), 201–226.
Sadler, M.J., 2016. Dried fruit and dental health. Int. J. Food Sci. Nutr. 67 (8), 944–959.
Toumba, K.J., Duggal, M.S., 1999. A single blind, six-way crossover healthy volunteer study to investigate the effect of two new formulations of blackcurrant drink in comparison to two conventional marketed fruit drinks, and two control treatments on the pH of dental plaque. Br. Dent. J. 186, 626–629.
West, N.X., Hughes, J.A., Parker, D.M., Moohan, M., Addy, M., 2003. Development of low erosive carbonated fruit drinks 2. Evaluation of an experimental carbonated blackcurrant drink compared to a conventional carbonated drink. J. Dent. 31, 361–365.
West, N.X., Hughes, J.A., Parker, D.M., Newcombe, R.G., Addy, M., 1999. Development and evaluation of a low erosive blackcurrant juice drink 2. Comparison with a conventional blackcurrant juice drink and orange juice. J. Dent. 27, 341–344.
West, N.X., Hughes, J.A., Parker, D., Weaver, L.J., Moohan, M., De'Ath, J., Addy, M., 2004. Modification of soft drinks with xanthan gum to minimise erosion: a study in situ. Br. Dent. J. 196, 478–481.

Further Reading

Issa, A.I., Toumba, K.J., Preston, A.J., Duggal, M.S., 2011. Comparison of the effects of whole and juiced fruits and vegetables on enamel demineralisation in situ. Caries Res. 45 (5), 448–452.
PHE, 2016. The Eatwell Guide Booklet—Helping You Eat a Healthy Balanced Diet. PHE, England.
Van Loveren, C., 2009. Oral and Dental Health: Prevention of Dental Caries, Erosion, Gingivitis and Periodontitis. ILSI, Europe, Belgium.
WHO, 2015. Guideline: Sugars Intake for Adults and Children. World Health Organization, Geneva, Switzerland.

Authorised EU health claim for sugar beet fibre

8

Janice I. Harland†
Harland Hall Associates, Poulton, Cirencester, United Kingdom

8.1 Introduction

The beet plant has been cultivated on Mediterranean shores since ancient times; the first reference dates to 420 BC (Winner, 1993). Initially it was grown mainly for its leaves which probably resemble modern-day Swiss chard. From around AD 800 there is a reference to the plant in Northern Europe, and by the end of the 15th century the crop was grown all over Europe. It appears that shortly thereafter the plant is reported as being valuable for the sweet-tasting juice from its root. By the 17th century, beet was regularly field cultivated, but was red or deep coloured and primarily used for cattle fodder. White-fleshed beet originated from Central Europe. The development of beet as an industrial crop did not start until the second half of the 18th century, with the first sugar beet factory erected in 1801–02 in Silesia, Prussia. The Napoleonic wars accelerated the growth of commercial sugar beet factories throughout Europe as the export trade lines (for sugar derived from cane) from the West Indies were disrupted. The commercially cultivated form of the sugar beet—*Beta vulgaris*—was thought to be named around 1830.

The fibre component of the sugar beet derived during commercial processing of the crop was traditionally used as animal feed (Harland, 1993a); its development as a fibrous food ingredient commenced during the 1980s in the United Kingdom as the ingredient Beta Fibre and in Sweden as Fibrex. The physiological effects of consuming Beta Fibre were first reviewed over 20 years ago (Harland, 1993b). In this review the first clinical study documented and evaluated the effect of sugar beet pulp on blood glucose and insulin levels (Tredger et al., 1981), and insights into the fermentability of sugar beet fibre (Nyman and Asp, 1982) and its effect in constipated patients (Giacosa et al., 1990) were also reported.

The importance of fibres of plant origin and the identification of the role that dietary fibre has on digestive health, specifically faecal output and intestinal transit times, became topical once again around 50 years ago, largely as a result of the epidemiological observations by Burkitt (1973) and Burkitt et al. (1972). The specific beneficial role of dietary fibre in faecal bulking and gut transit was later endorsed in an extensive review of the literature relating to the effect of dietary fibre on faecal weight and composition (Cummings, 1993). A further paper by Cummings et al. (2004), conducted as part of the Process for the Assessment of Scientific Support for Claims on Foods (PASSCLAIM) initiative, identified the specific measures of digestive health that could potentially be used to support a health claim related to gut health and immunity. Latterly

† Deceased

Foods, Nutrients and Food Ingredients with Authorised EU Health Claims. http://dx.doi.org/10.1016/B978-0-08-100922-2.00008-5
Copyright © 2018 Elsevier Ltd. All rights reserved.

specific advice on providing scientific substantiation of these claims has been provided by EFSA [EFSA Panel on Dietetic Products Nutrition and Allergies (NDA), 2011a]; this guidance was reviewed following a period of consultation (EFSA, 2014).

Many of the Article 13 claims relating to gut health rely on studies published in the literature, thus Cummings' (1993) review provides a good insight into our current knowledge about vegetable fibres and faecal bulking, which when supplemented by two studies specific to sugar beet fibre, provides the scientific substantiation for this claim.

8.2 Characterisation of the ingredient

The subject of the claim is sugar beet fibre which includes the fibre of all species within *B. vulgaris* L. The fibrous ingredient is characterised by a content of hemi-celluloses (22%–32%), pectins (22%–29%) and cellulose (19%–28%), with little or no lignin (<3%) (Thibault et al., 2001). Typically the fibre content of sugar beet fibre is between 70 and 80 g/100 g, with a limited content of protein (5%), ash (3%) and moisture (7%). The ratio of insoluble to soluble fibre is approximately 2:1. Sugar beet fibre does not have a conventional hemi-cellulose composition, instead the main cell wall polysaccharide is >90% arabinose polymers. These are primarily arabinans with an arabinogalactan side chain which links to the pectin and forms a small part of the hemi-cellulose. The sugar beet cell walls have a high content of pectin which derives from galacturonic acid. The pectin backbone is α-(1→4)-linked D-galacturonic acid interspersed with a few α-(1→2)-linked rhamnose in the ratio >10:1. Pectin in sugar beet fibre has a high degree of methylation although lower than citrus or apple pectins, a relatively low viscosity and it does not form gels under the usual conditions. Cellulose is usually the second most abundant polymer and is largely present as type I with limited lateral order. The phenolic component in sugar beet fibre is largely present as ferulic acid (<10 mg/g).

The two main commercial sources of sugar beet fibre are Fibrex and Beta Fibre. The manufacturing process is marginally different for these two sources of the fibre, but the nutrient composition is remarkably similar, largely because the pectin component is relatively stable and the extraction and manufacturing process has little effect on it.

The applicant of the Article 13.5 claim markets Fibrex in the form of different powder grains with different particle sizes, coarse (non-milled) and coarse grains pressed into flakes [EFSA Panel on Dietetic Products Nutrition and Allergies (NDA), 2011b].

The subject of the claim is sugar beet fibre present in foods and those forms added to foods; the EFSA NDA Panel considered that the food constituent sugar beet fibre is sufficiently characterised relative to the claimed effect [EFSA Panel on Dietetic Products Nutrition and Allergies (NDA), 2011b].

8.3 Authorised Article 13.5 claim

There are few authorised Article 13.5 claims and it is worth reviewing the definition of this group of claims which states,'...Any additions of claims to the list referred to in paragraph 3 based on newly developed scientific evidence and/or which include a request for the protection of proprietary data...'(2006). The term 'newly developed scientific evidence' is an interesting term and does not mean new scientific evidence, as will become evident, although it has often been wrongly interpreted as such. In essence it allows any data from any era to be reassessed and collated in a manner that would be appropriate to provide scientific substantiation in support of a health claim.

The authorised Article 13.5 claim is that, 'Sugar beet fibre contributes to an increase in faecal bulk'. The conditions of use to bear the claim require that a food should be at least, 'high in fibre' as per the Annex to Regulation (EC) No. 1924/2006. Although the applicant suggested that a consumption pattern of over 7 g/day was required, this was not included in the conditions of use. The claimed effect is 'contributes to an increase in faecal bulk' and the proposed target population for the health claim comprises people who want to improve or maintain a normal bowel function. EFSA has established that increasing faecal bulk may be a beneficial physiological effect [EFSA Panel on Dietetic Products Nutrition and Allergies (NDA), 2011a]. In its opinion, EFSA stated that a cause-and-effect relationship has been established between the consumption of sugar beet fibre and the increase in faecal bulk, and that the wording reflects the scientific evidence: 'sugar beet fibre increases faecal bulk'. Although the EFSA opinion was published in 2011, it was January 2014 before the EU Regulation authorising the claim was published in the Official Journal (EU, 2014) indicating the time it can take to get a claim embedded into the Register of Claims.

8.3.1 Increased faecal bulk: some general considerations

The measurement of faecal bulk is central to any claim relating to gut health, as is an appropriate definition of fibre. While fibre was initially considered to be effective, as it provided simple bulking, subsequently described as water-holding capacity, it is now recognised that the physiochemical properties of the fibre are important and help to explain the differences in response in faecal output to various fibres.

When considering the physical properties of fibre, particle size is an important consideration. For the same fibre fed under the same conditions, generally the greater the particle size, the larger the effect on faecal output. It is believed that large particles present a smaller net surface area and thus are more slowly degraded in the gut and more likely to escape degradation. As a result, they provide bulk to the digesta and are able to exert a direct physical effect on the colon and a surface for fermentation by the bacteria (Cummings, 1993). The water uptake of sugar beet fibre is high as compared to cereal fibres (Thibault et al., 2001). Hydration capacity—swelling—is determined by the water retention capacity (WRC) and the water absorption capacity (WAC). Both of these factors are high in sugar beet fibre, thus the physical particle size of sugar beet fibre is less critical than it would be for, say, a cereal fibre, as the functional particle size within the digesta is likely to be some four- to eightfold greater than the dry particle (Nordic Sugar, 2014).

Details of the techniques used to measure faecal bulk were not given in the EFSA guidance, other than reference to 'generally accepted methods'. The PASSCLAIM initiative not only evaluated suitable markers and their measurement for the scientific substantiation of gut health and immunity claims (Cummings et al., 2004), but also defined normal bowel habit and how it may be measured. For example in respect of faecal output, it identified the gold standard method as the use of single or multiple doses of radio-opaque pellets.

The methods of analysis of fibre have changed regularly through the years; fibre is currently defined within the European Union by Regulation (EU, 2011) as being carbohydrate polymers with three or more monomeric units which are neither digested nor absorbed in the small intestine of human and belong to the following categories:

- edible carbohydrate polymers naturally occurring in food as consumed;
- edible carbohydrate polymers, which have been obtained from food raw material by physical, enzymatic or chemical means and which have a beneficial physiological effect demonstrated by generally accepted scientific evidence and
- edible synthetic carbohydrate polymers which have a beneficial physiological effect demonstrated by generally accepted scientific evidence.

This definition excludes lignin, as it is not a carbohydrate polymer. However, lignin often plays a role in rendering the carbohydrate polymers less accessible to degradation and may be important in modifying faecal output. It is therefore deemed appropriate that the dietary fibre measurement techniques, either AOAC or AACC, capture lignin in the analysis and hence it is usually included in the fibre value as declared. As sugar beet fibre contains a very low level of lignin, the method of analysis used tends to produce less variable results. Consequently the non-starch polysaccharide (NSP) content of sugar beet fibre and its fibre content determined by dietary fibre analysis are similar at c. 73%–76%. However the AOAC analysis can yield a slightly lower value of approx. 67% due to the loss of some of the smaller molecules (Nordic Sugar, 2014).

8.3.2 Summary of substantiating evidence

The first study with sugar beet pulp cited by Cummings dates from 1936 (Williams and Ormsted), where 27 g of sugar beet fibre intake was associated with a 67 g increase in faecal weight (Cummings, 1993). The apparent increase in faecal weight per gram of dietary fibre fed was calculated to be 4.9 g/g and comparable to wheat bran fibre (5.4 g/g ± SEM 0.7 g) (Cummings, 1993).

The applicant identified four human studies pertinent to the health claim. However EFSA considered that as faecal bulk was not measured in two studies, they could not draw any conclusion from these studies. One of these studies was conducted in people with chronic constipation and measured stool frequency (Giacosa et al., 1990), while the second studied the effects of sugar beet fibre on cholesterol, bile acid and total ileostomy excretion in nine ileostomy patients (Langkilde et al., 1993).

The two studies used in scientific substantiation were a randomised, cross-over study that evaluated the effect of supplemental sugar beet and wheat fibre on faecal weight and intestinal transit time in 17 healthy men (20–40 years) (Lampe et al., 1993) and a Latin-square design randomised study (3 × 3), in 9 healthy men (mean age 21.5 years) that consumed either a control diet (22 g/day dietary fibre) or the same control diet supplemented with 50 g/day of either sugar beet fibre or inulin from chicory for 28 days each (Castiglia-Delavaud et al., 1998). Thus the data to substantiate the claim targeted at the population as a whole is provided by the results obtained in 26 healthy men aged 18–40 years.

In the first of these studies, subjects consumed 20 g/day of either sugar beet fibre (DuoFiber) or wheat fibre, ground to a similar particle size distribution, for 30 days each after a 20-day run-in period (Lampe et al., 1993). There was no wash-out period between treatments. Subjects were asked to maintain their usual diet for the duration of the study and the fibre supplements were consumed with meals. Faeces were collected from day 11 to day 20 of the run-in and during each treatment period. Of the 20 commencing the study, 17 men completed and entered data analysis. EFSA noted that while the effect of treatment was assessed by analysis of variance (ANOVA), the full statistical methodology was not explained. Wet and dry faecal weights during consumption of sugar beet fibre were significantly higher than during the run-in (self-selected diet) period (mean ± SEM; wet faecal weight: 253 ± 10 vs. 198 ± 10 g/day, $P < 0.0001$; dry faecal weight: 57 ± 2 vs. 49 ± 2 g/day, $P < 0.0001$ respectively). From these data, it can be calculated that each gram of total dietary fibre from sugar beet fibre was associated with an increase in wet and dry faecal weight of 2.75 and 0.4 g/day respectively. No significant differences were reported in relation to increased faecal weight between the sugar beet and wheat bran fibre groups, but there was an increase in faecal wet weight with sugar beet fibre and faecal dry weight with wheat bran. The EFSA Panel notes that wheat fibre has an established role in increasing faecal bulk in humans [EFSA Panel on Dietetic Products Nutrition and Allergies (NDA), 2010]. The Panel considered that this study shows an effect of sugar beet fibre on increasing faecal bulk.

In the second, Latin-square, study (Castiglia-Delavaud et al., 1998) the control diet was compared with sugar beet fibre (source not given) and inulin. Each experimental period comprised 2 days (days 1 and 2) with the control diet, 14 days (days 3–16) with a progressive adaptation to the intervention up to a maximum of 50 g/day fibrous ingredient (sugar beet fibre 71.0 g/kg NSP + lignin) and inulin (62 g/kg inulin) and 12 days (days 17–28) with a constant maximum intake of fibre. Diets were partially controlled (lunch and dinner were provided by the nutrition laboratory). Primary outcomes of the study were metabolisable energy and net metabolisable energy of the diets. Secondary outcomes were energy expenditure, faecal weight, frequency of defecations and faecal bacterial mass. The Panel notes that faecal weight was a secondary endpoint of the study. Comparisons between experimental diets were performed by ANOVA using the general linear models procedure according to a Latin-square design with three repetitions. Wet and dry faecal weights during consumption of sugar beet fibre were significantly higher than during the control diet (mean ± SEM; wet faecal weight: 202 ± 16 vs. 129 ± 16 g/day, $P < 0.05$; dry faecal weight: 37.6 ± 2.3 vs. 27.7 ± 2.3 g/day, $P < 0.05$, respectively).

It can be calculated that each gram increase in sugar beet fibre consumption (NSP + lignin) was associated with an increase in wet and dry faecal weight of approx. 2.5 and 0.35 g/day respectively. Mean faecal bacterial weight (presented as dry matter) was significantly higher in the sugar beet fibre group in comparison with the control diet (20.3 ± 1.2 vs. 13.0 ± 1.2 g/day, $P < 0.01$). No significant differences were reported in relation to wet or dry faecal weights between the sugar beet and inulin fibres.

The Panel considers that this study shows an effect of sugar beet fibre on increasing faecal bulk. The Latin-square study (Castiglia-Delavaud et al., 1998) also provides an insight into the mechanism, by providing information about the extent of sugar beet NSP fibre digestion which is extensive at 0.825 (SD 0.088) and microbial synthesis which was 56% higher when sugar beet was included in the diet.

There is no doubt that both these studies and the earlier study cited (Cummings, 1993) demonstrate an increase in faecal weight, although the magnitude of the increase reported in the most recently conducted studies indicate a lesser effect on faecal weight compared with that reported by Cummings, for example 2.5–2.75 versus 4.9 g/g sugar beet fibre. This difference may be due to the different methods of fibre analysis used or the different sources of sugar beet fibre, or the wide variation in faecal output reported in both of the more recent studies. It is generally assumed that the inclusion of 1 g dietary fibre will increase faecal output by around 4 g/day (Southgate, 1990) and sugar beet fibre is close to this value.

Unusually, also considered in the scientific substantiation for this claim were three animal studies; these evaluated the effects of sugar beet fibre on faecal bulk (Johnson et al., 1990; Klopfenstein, 1990; Nyman and Asp, 1982). In two of these studies, addition of sugar beet fibre (100 g/kg feed) to the usual diet (Nyman and Asp, 1982) or to a fibre-free diet (Johnson et al., 1990) induced a significant increase in faecal wet and dry weights in rats. In the third study (Klopfenstein, 1990), coarse and fine sugar beet fibre fed at different doses (5, 7.5 and 10% of the diet) induced a dose–response increase in wet and dry faecal weights, which was independent of particle size. The Panel considered that the evidence provided by these three animal studies supported an effect of sugar beet fibre on increasing faecal bulk.

8.3.3 Mechanisms of action

While wet and dry faecal weights were increased during consumption of sugar beet fibre compared with control, no significant differences were reported in relation to increased faecal weight between the sugar beet and wheat bran fibres (Lampe et al., 1993). However the nature of the increase in faecal weight was quite different between the two fibres with sugar beet fibre increasing faecal wet weight and wheat bran increasing faecal dry weight. The authors concluded it was uncertain whether the higher moisture content in the faeces was due to increased water-holding capacity in the residual sugar beet fibre or increased bacterial water as a result of its fermentation. The EFSA Panel explained the mechanisms of action of sugar beet fibre as being a result of the insoluble components of sugar beet fibre increasing faecal bulk by absorbing water in the large intestine and the soluble components being fermented by bacteria in the large intestine leading to an increase in bacterial mass [EFSA Panel on Dietetic Products Nutrition and Allergies (NDA), 2011b]. The high digestibility proposed for sugar beet fibre and the low level of lignification would suggest that there is little of the fibre left unfermented (Castiglia-Delavaud et al., 1998). Extensive utilisation of the fibre was also reported when sugar beet fibre was compared with wheat bran in an in vitro semi-continuous culture fermentator (Rumney et al., 1990). In this study with sugar beet fibre, there was a much greater volatile fatty acid production than with wheat bran, indicating that the sugar beet dietary fibre was more extensively used. The additional volatile fatty acid was largely due to an increased production of acetate.

The high water uptake properties of sugar beet fibre may be relevant with regard to rate of passage of the digesta, increasing the functional mass of the particles, but the animal study (Klopfenstein, 1990) included in the scientific substantiation reported no difference in faecal bulking related to particle size, possibly as a result of these water uptake properties. Thus whether water uptake has a substantial effect in faecal bulking remains to be proven. The clinical study provides evidence of substantial bacterial synthesis and it may be this mass and the higher water content of bacteria that is of importance in adding to the faecal bulk (Castiglia-Delavaud et al., 1998).

8.4 Article 13.1 claim

Health claim applications for sugar beet fibre were also made as part of the call for claims in the Article 13.1 category [Regulation (EC) No. 1924/2006]. The submitted claims relate to a reduction of post-prandial glycaemic responses, maintenance of normal blood glucose concentrations and the maintenance of normal blood low-density lipoprotein (LDL)-cholesterol concentrations [EFSA Panel on Dietetic Products Nutrition and Allergies (NDA), 2011c]. Of particular interest is the claim relating to the maintenance of normal blood LDL-cholesterol concentrations. This opinion included three references to substantiate the claim (Cossack and Musaiger, 1991; Hagander et al., 1989; Israelsson et al., 1993). The first study refers to a placebo-controlled, randomised cross-over study with 12 patients with type 2 diabetes that consumed 40 g sugar beet fibre/day or placebo for 8 weeks each. Both the control and treatment groups reduced LDL-cholesterol and the Panel commented on the lack of effect and small number of subjects. The second trial reported on 10 desert nomads who, after a baseline period of 2 weeks, consumed 36 g sugar beet fibre/day for 5 weeks, followed by a second baseline period of 4 weeks. The Panel considered that no conclusions could be drawn from this uncontrolled study (Cossack and Musaiger, 1991). The third study (Israelsson et al., 1993) conducted with Fibrex was a placebo-controlled, cross-over study in 30 hypercholesterolaemic women who consumed 30 g sugar beet fibre/day or placebo for 4 weeks each, in a random order after a 1-month run-in period in the context of a moderate low-fat, low-cholesterol diet. Sugar beet and placebo supplements (containing 3.5% fibre by weight) were administered either as crumbs or as bread in the context of meals in 5 g portions. A significant reduction in LDL-cholesterol concentrations (−6.5%) and in the LDL/high-density lipoprotein (HDL) cholesterol ratio (−13.5%) was observed during the sugar beet fibre intervention compared with placebo.

In respect of this finding, the Panel stated that as sugar beet fibre contains 22%–29% pectins by weight, the blood cholesterol-lowering effect observed when 30 g of sugar beet fibre/day was consumed (containing the equivalent of about 6 g pectins/day), was in the range of what could be expected from the consumption of 6 g pectin/day. It was further assumed that the effect on cholesterol was dependant on the water-soluble fibres—essentially the effect of pectins—and likely to be dependent on its viscosity, which reduces the reabsorption of bile acids, increases the synthesis of bile acids from cholesterol and reduces circulating blood cholesterol concentrations. It was assumed that as the remaining fibre components in sugar beet fibre (hemi-cellulose and cellulose) are non-viscous and insoluble, they are not expected to exert a cholesterol-lowering effect. Thus the Panel concluded that a cause-and-effect relationship was not established between the consumption of sugar beet fibre and maintenance of normal blood LDL-cholesterol concentrations *beyond the hypocholesterolaemic effects which could be expected from the pectin content* of sugar beet fibre, and further remarked that a claim on pectins and maintenance of normal blood cholesterol concentrations has already been assessed with a favourable outcome [EFSA Panel on Dietetic Products Nutrition and Allergies (NDA), 2010, 2011c]. It is interesting that the EFSA Panel considered pectins to be sufficiently characterised. It is reported in the pectin opinion that pectins are composed of linear chains of α-(1,4)-galacturonic acid units with varying degrees of methylation and side chains including galacturonic and glucuronic acids [EFSA Panel on Dietetic Products Nutrition and Allergies (NDA), 2010]. In fact while the pectin backbone of sugar beet pectin is α-(1→4)-linked D-galacturonic acid, it is interspersed with a few α-(1→2)-linked rhamnose in a ratio >10:1, not galacturonic and glucuronic acids (Thibault et al., 2001). While the degree of methylation in sugar beet pectin is high, it is much lower than citrus or apple pectins and has very different gelling properties. The scientific substantiation identified in the opinion relies largely on a meta-analysis conducted with soluble fibres (Brown et al., 1999). From this meta-analysis it was estimated that 1 g pectins/day produced a change in total and LDL-cholesterol concentrations of −0.07 (95% CI, −0.117 to −0.022) mmol/L and −0.05 (95% CI,−0.087 to −0.022) mmol/L respectively. Thus 6 g pectin/day would be anticipated to reduce LDL cholesterol by c. 0.42 mmol/L, whereas the reduction reported for sugar beet fibre was c. 0.22 mmol/L (Israelsson et al., 1993).

In a further study with sugar beet fibre not cited in the EFSA opinion, the effect of source of dietary fibre supplementation on blood lipids was assessed in 12 normocholesterolemic men (Tredger et al., 1991). Guar gum, sugar beet fibre (Beta Fibre) or wheat bran was incorporated into bread with half a loaf a day, providing 20 g of the fibre preparation. Each bread was eaten for 14 days with a 21-day wash-out period between each stage of the study. Total cholesterol levels were reduced by an average of 11.7% ($P < 0.01$) during guar supplementation and 4.6% ($P < 0.025$) during sugar beet fibre supplementation, while bran supplementation had no effect on blood lipids. In a second study conducted by the same research team in seven type 2 diabetics consuming 18-g dietary fibre from Beta Fibre, reductions in total cholesterol, triacylglycerol (TAG) and apolipoprotein B of 6.2, 10.6 and 6.0%, respectively, and all $P < 0.05$ were reported when compared with the control blood lipid levels (Travis et al., 1990).

In all of the identified studies with sugar beet fibre, reductions of total cholesterol have been reported. These range from 3.6% to 7.8% which is typical of the level of reduction in cholesterol reported for this type of dietary intervention. Although many of the identified studies are small, the totality of the evidence supports a consistent reduction of total and LDL-cholesterol. There are also a number of animal studies that support this finding (Harland, 1993b). Whether the mechanism of action is solely due to the pectin in sugar beet fibre remains to be proven. A pectic hemi-cellulosic fraction of sugar beet fibre was reported to reduce cholesterol by 54% in rats (Belsten et al., 1994), while a further rat study suggested a mechanism involving disruption of the enterohepatic bile acid circulation, possibly via changes in the rate of absorption of dietary lipid (Overton et al., 1994).

Thus a health claim can be made for sugar beet fibre and the maintenance of normal cholesterol, but the claim must relate to the pectin content of sugar beet fibre, thus 'Consumption of the pectins in sugar beet fibre contributes to the maintenance of normal blood cholesterol levels'.

To bear the claim, at least 6 g pectin/day is required. This is equivalent to 25–30 g sugar beet fibre.

8.5 Other legislation

Within the European Union the main other relevant legislation is the provision of food information to consumers that aims to ensure that labels are clear and understandable and to assist consumers wishing to make better-informed food and dietary choices (EU, 2011). This Regulation prescribes the format of the nutrition declaration and also makes the provision for additional nutrition information on a separate face (e.g. front of pack labelling), provided it does not distract from that information which is mandatory; this mandatory information should all be in the same field of vision. However, Regulation (EU) No. 1169/2011 does not provide a reference intake for fibre and it is no longer compulsory to make a fibre declaration within the standard nutrition declaration. If a claim is made for fibre, then it is mandatory to include it within the nutrition declaration and there is also the provision that fibre may continue to be declared on a voluntary basis.

Advice on the application of this Regulation is available from the UK Government (https://www.food.gov.uk/sites/default/files/fir-guidance2014.pdf).

Within the USA a heart disease claim similar to the Article 13.1 cholesterol management claim can be made for sugar beet fibre (2012); it is generic in nature and relates to 'fruits, vegetables, and grain products that contain fiber, particularly soluble fiber and risk of coronary heart disease'. The claim states that, 'diets low in saturated fat and cholesterol and high in fruits, vegetables, and grain products that contain fiber "may" or "might" reduce the risk of heart disease'. The claim is limited to those fruits, vegetables and grains that contain fibre, provided the food contains, without fortification, at least 0.6 g of soluble fibre per reference amount customarily consumed.

8.6 Conditions of use, warning labels and safety issues

To carry the claim, 'Sugar beet fibre contributes to an increase in faecal bulk', a food should be at least 'high in fibre' as described in the Annex to Regulation which is defined as being at least 6 g fibre/100 g or 3 g fibre/100 kcal [Regulation (EC) No. 1924/2006]. The target population is the general population.

The claim relates to pectin and the maintenance of normal blood cholesterol concentrations. To bear the claim, at least 6 g pectins/day should be consumed in one or more servings. The target population is adults.

Neither are there any warning labels that are recommended nor any specific safety issues related to the use of the health claims.

8.7 Consumer issues

Most of the chronic diseases and many bowel disorders may be attributed to lack of faecal bulk (Southgate, 1990). Thus the population at large could benefit from the inclusion of more fibre in the diet, but it has been suggested that extracted fibres may not have the same physiological functionality as foods naturally rich in fibre (SACN, 2008) and evidence related to their functionality is limited (SACN, 2015). Sugar beet fibre is not extracted but concentrated by the removal of sugar from the beet. Hence sugar beet fibre, as a concentrated source of fibre with interesting physiological properties, may help meet the demand for a greater fibre intake in the diet and falls within the definition of fibre as 'a variety of food sources' proposed by SACN (2015).

The 2015 SACN report reevaluates the UK dietary reference value for the average population intake of dietary fibre for UK adults and suggests that this should be 30 g/day (fibre as defined and measured using the AOAC methods). The previous UK dietary reference value of 18 g/day NSP, defined by the Englyst method, equates to about 23–24 g/day of dietary fibre if analysed using AOAC methods, hence the new recommendation represents an increase of approximately 6 g or 25% from the original value, a recommendation achieved by a relatively small proportion (19%) of the adult population (Harland, 2012). Dietary fibre intakes for UK children were also recommended and these were 15 g/day for 2–5 year olds, 20 g/day for 5–11 year olds, 25 g/day for 11–16 year olds and 30 g/day for adolescents aged 16–18 years (SACN, 2015). The new UK dietary reference values are based on evidence that the consumption of a variety of foods rich in dietary fibre as a naturally integrated component is associated with beneficial health outcomes. It is also suggested that a diet rich in these foods will also usually be rich in micronutrients and phytochemicals that may have additional health benefits. Currently the main sources of fibre in the UK diet are cereals and cereal products, which provide about 40% of intake across age groups, and vegetables and potatoes (a quarter to a third of intake) (SACN, 2015).

Various beneficial physiological effects for diets rich in natural dietary fibre components are suggested such as increasing stool bulk, decreasing intestinal transit time and constipation, or lowering total and LDL-cholesterol as well as growing evidence of a reduced risk of type 2 diabetes mellitus, cardiovascular disease and colo-rectal cancer (SACN, 2015).

Among United Kingdom and many other European consumers, few achieve the current fibre recommendations and in the United Kingdom there is a now a challengingly higher intake to achieve which many consumers will find difficult in practice.

Consumers may not be aware of the health benefits of fibre beyond laxation, and the motivation to increase fibre content of the diet may just simply not be present. Thus while there is a growing need to increase fibre content of the diet, a commitment to change eating habits is required to achieve this, and it is necessary to start at breakfast by consuming high-fibre bread and breakfast cereals, maintaining a 3–4 g increased intake at each meal and preferably at each snacking occasion and consuming eight portions of fruit and vegetables per day (Stanner, 2015). However, there is a paucity of high-fibre health snacks and this is an area of opportunity for a fibre-rich ingredient such as sugar beet fibre.

The wording of the claim, sugar beet fibre contributes to an increase in faecal bulk, as authorised for use in the European Union is not particularly appealing to consumers. Use of expressions such as 'improving digestive health' or 'good for your digestive system/bowels' may be marginally more acceptable but care has to be taken that any alteration in the wording does not strengthen the claim.

Other consumer considerations are that sugar beet fibre is allergen-free and is naturally free from gluten which means it is an excellent alternative dietary fibre source for people suffering from gluten intolerance. In addition, unlike cereal products such as bran and flour, it contains no phytic acid which can form complexes with divalent minerals such as iron and zinc that can result in reduced bioavailability and consequently impairment or reduction in the natural absorption of these essential minerals in the body.

8.8 How the claims are being used in practice and future trends

The manufacturers of the commercially available food ingredient, Fibrex, anticipate that the claim provides new marketing opportunities for producers of bakery products, cereals, cereal bars, fruity chewy bars etc. It is envisaged that sugar beet fibre could be used alone or in combination with other fibrous food ingredients that can carry a similar claim.

Another property of sugar beet fibre is its high water absorption property; it will take up to 3–4 times its weight in water, resulting in a further reduction in its already low-energy density when it takes up water. In essence it acts like a sponge and will retain much of the water it takes up, making it a useful ingredient for low-energy products.

Manufacturers producing gluten-free ranges have identified sugar beet fibre as a useful ingredient in their recipes. Artisan bakers and others manufacturing gluten-free bread and bakery products use sugar beet fibre and comment that the resulting bread is softer and fresher as a result of the moisture that it retains within the dough.

Sugar beet fibre has been used to enhance the quality of Scandinavian ready meals, on trial in UK supermarkets. The traditional Swedish meatball has been given a makeover to exclude gluten and make it suitable as a gluten-free ready meal. The water-binding properties have also been identified to enhance the juiciness of meat products.

Other meat products that have been enhanced by the inclusion of sugar beet fibre are a high-fibre sausage that includes sugar beet fibre and meets the requirement for a source of fibre claim (3 g fibre/100 g) [Regulation (EC) No. 1924/2006], beef burgers, frankfurters and other processed meats.

Sugar beet fibre has also been identified as being suitable for use in coatings for meat and chicken, in savoury snacks and soups. However a search of the supermarket shelves, at the time of writing, revealed few products carrying the sugar beet fibre claim.

8.9 Conclusions

In the European Union, there is an authorised Article 13.5 health claim relating to sugar beet fibre and contributes to an increase in faecal bulk. For foods to carry the claim, they must be high-fibre foods as defined in the Annex to the Regulation (EC) No. 1924/2006 and contain at least 6 g fibre/100 g or 3 g fibre/100 kcal. The scientific substantiation of the claim is mainly from two small clinical studies in men; no studies were considered that were conducted in women and three rat studies were also considered.

No information is given to suggest that there is a dose–response relationship in humans, although a rat study is cited as evidence and also a meta-analysis relating to soluble fibres more generally (Klopfenstein, 1990). There is no indication that there is a threshold for the effect.

It is proposed that the mechanisms of action of sugar beet fibre are twofold; the insoluble components of sugar beet fibre increases faecal bulk by absorbing water in the large intestine and the soluble components are fermented by bacteria in the large intestine, leading to an increase in bacterial mass [EFSA Panel on Dietetic Products Nutrition and Allergies (NDA), 2011b]; although disruption of the enterohepatic bile acid circulation, may also be relevant (Overton et al., 1994).

A second claim may be made for sugar beet fibre relating to its pectin content. An Article 13.1 generic claim exists for pectins including those from sugar beet and the maintenance of normal cholesterol, thus it can be claimed that, 'Consumption of the pectins in sugar beet fibre contributes to the maintenance of normal blood cholesterol levels'. To bear the claim, at least 6 g pectin/day is required. This is equivalent to 25–30 g sugar beet fibre.

8.10 Sources of further information and advice

The EFSA website provides a register of questions that relate to all the Article 13 claims that have been asked of them (http://registerofquestions.efsa.europa.eu/roq-Frontend/questionsListLoader?panel=ALL).

It also provides, in its journal, guidance related to decisions made on claim applications and guidance for making claims (http://www.efsa.europa.eu/en/publications/efsajournal.htm).

Guidance on Nutrition and Health Claims Regulation can also be found on the European Commission website (http://ec.europa.eu/food/food/labellingnutrition/claims/index_en.htm). Information about Regulation 1169/2012 can be found at: https://www.food.gov.uk/sites/default/files/fir-guidance2014.pdf.

Guidance of methods of fibre analysis can be found at: http://ec.europa.eu/food/food/labellingnutrition/nutritionlabel/guidance_methods_analysis_fibre_dec2012.pdf.

Information about the dietary fibre European DRV can be found in: EFSA Panel on Dietetic Products, Nutrition, and Allergies (NDA), 2010. Scientific Opinion on dietary reference values for carbohydrates and dietary fibre. EFSA J. 8 (3), 77. Available from: www.efsa.europa.eu

The SACN report provides the most up-to-date summary of the science relating to nutrition, health and dietary fibre and the UK DRV. Available from the UK Government websites (https://www.gov.uk/government/uploads/system/uploads/attachment_data/file/445503/SACN_Carbohydrates_and_Health.pdf and https://www.gov.uk/government/uploads/system/uploads/attachment_data/file/339367/SACN_Draft_position_statement_on_dietary_fibre_and_health_and_dietary_fibre_definition_2008.pdf).

References

Belsten, J.L., Brown, J.C., Holden, S., Majsak-Newman, G., Smith, T., Livesey, G., 1994. Hypocholesterolaemic non-starch polysaccharide from sugar beet. Int. J. Food Sci. Nutr. 45, 49–61.

Brown, L., Rosner, B., Willett, W.W., Sacks, F.M., 1999. Cholesterol-lowering effects of dietary fiber: a meta-analysis. Am. J. Clin. Nutr. 69, 30–42.

Burkitt, D.P., 1973. Epidemiology of large bowel disease: the role of fibre. Proc. Nutr. Soc. 32, 145–149.

Burkitt, D.P., Walker, A.R., Painter, N.S., 1972. Effect of dietary fibre on stools and the transit-times, and its role in the causation of disease. Lancet 2, 1408–1412.

Castiglia-Delavaud, C., Verdier, E., Besle, J.M., Vernet, J., Boirie, Y., Beaufrere, B., De Baynast, R., Vermorel, M., 1998. Net energy value of non-starch polysaccharide isolates (sugarbeet fibre and commercial inulin) and their impact on nutrient digestive utilization in healthy human subjects. Br. J. Nutr. 80, 343–352.

Cossack, Z.T., Musaiger, A.O., 1991. Effect on lipid metabolism of beet fibre in desert nomads with low habitual fibre intake. Eur. J. Clin. Nutr. 45, 105–110.

Cummings, J.H., 1993. The effect of dietary fiber on fecal weight and composition. CRC Handbook of Dietary Fiber in Human Nutrition. second ed. CRC Press Inc., Boca-Raton, Florida, USA, pp. 263–349.

Cummings, J.H., Antoine, J.M., Azpiroz, F., Bourdet-Sicard, R., Brandtzaeg, P., Calder, P.C., Gibson, G.R., Guarner, F., Isolauri, E., Pannemans, D., Shortt, C., Tuijtelaars, S., Watzl, B., 2004. PASSCLAIM: gut health and immunity. Eur. J. Nutr. 43 (Suppl. 2), II118–II173.

EFSA Panel on Dietetic Products Nutrition and Allergies (NDA), 2010. Scientific Opinion on the substantiation of health claims related to pectins and reduction of post-prandial glycaemic responses (ID 786), maintenance of normal blood cholesterol concentrations (ID 818) and increase in satiety leading to a reduction in energy intake (ID 4692) pursuant to Article 13(1) of Regulation (EC) No. 1924/2006. EFSA J. 8 (10), 1747.

EFSA Panel on Dietetic Products Nutrition and Allergies (NDA), 2011a. Guidance on the scientific requirements for health claims related to gut and immune function. EFSA J. 9, 1984.

EFSA Panel on Dietetic Products Nutrition and Allergies (NDA), 2011b. Scientific Opinion on the substantiation of a health claim related to sugar beet fibre and increasing faecal bulk pursuant to Article 13(5) of Regulation (EC) No. 1924/2006. EFSA J. 9 (12), 2468.

EFSA Panel on Dietetic Products Nutrition and Allergies (NDA), 2011c. Scientific Opinion on the substantiation of health claims related to sugar beet fibre and reduction of post-prandial glycaemic responses (ID 752), maintenance of normal blood glucose concentrations (ID 752), maintenance of normal blood LDL-cholesterol concentrations (ID 809), and changes in bowel function (ID 751) pursuant to Article 13(1) of Regulation (EC) No. 1924/2006. EFSA J. 9 (4), 2034.

EU, 2011. Regulation (EU) No. 1169/2011 of the European Parliament and of the Council of 25 October 2011 on the provision of food information to consumers, amending Regulations (EC) No. 1924/2006 and (EC) No. 1925/2006 of the European Parliament and of the Council, and repealing Commission Directive 87/250/EEC, Council Directive 90/496/EEC, Commission Directive 1999/10/EC, Directive 2000/13/EC of the European Parliament and of the Council, Commission Directives 2002/67/EC and 2008/5/EC and Commission Regulation (EC) No. 608/2004. Off. J. Eur. Union L304, 18.

EU, 2014. Commission Regulation (EU) No. 40/2014 of 17 January 2014 authorising a health claim made on foods, other than those referring to the reduction of disease risk and to children's development and health and amending Regulation (EU) No. 432/2012. Off. J. Eur. Union L14, 8.

European Food Safety Authority, 2014. Discussion paper on the revision of the guidance on the scientific requirements for health claims related to gut and immune function. EFSA Supporting Publication 2014EN-NNNN. 6 p.

Giacosa, A., Sukkar, S., Frascio, F., Ferro, M., 1990. Sugar beet fibre: a clinical study in constipated patients. In: Southgate, A., Waldron, K., Johnson, I., Fenwick, G. (Eds.), Dietary Fibre: Chemical and Biological Aspects. AFRC Institute of Food Research, Norwich, pp. 355–361, (Special Publication No. 83).

Hagander, B., Asp, N.G., Ekman, R., Nilsson-Ehle, P., Schersten, B., 1989. Dietary fibre enrichment, blood pressure, lipoprotein profile and gut hormones in NIDDM patients. Eur. J. Clin. Nutr. 43, 35–44.

Harland, J.I., 1993a. 16. By-products. In: Cooke, D.A., Scott, R.K. (Eds.), The Sugar Beet Crop. Chapman and Hall, London, pp. 619–647.

Harland, J.I., 1993b. Beta Fibre: a case history. Int. J. Food Sci. Nutr. 44, S87–S96.

Harland, J.I., 2012. Food combinations for cholesterol lowering. Nutr. Res. Rev. 25, 249–266.

Israelsson, B., Jarnblad, G., Persson, K., 1993. Serum cholesterol reduced with Fibrex-a sugar beet fibre preparation. Scand. J. Nutr. 37, 146–149.

Johnson, I.T., Livesey, G., Gee, J.M., Brown, J.C., Wortley, G.M., 1990. The biological effects and digestible energy value of a sugar-beet fibre preparation in the rat. Br. J. Nutr. 64, 187–199.

Klopfenstein, C.F., 1990. Nutritional properties of coarse and fine sugar beet fiber and hard red wheat bran. I. Effects on rat serum and liver cholesterol and triglycerides and on fecal characteristics. Cereal Chem. 67, 538–541.

Lampe, J.W., Wetsch, R.F., Thompson, W.O., Slavin, J.L., 1993. Gastrointestinal effects of sugarbeet fiber and wheat bran in healthy men. Eur. J. Clin. Nutr. 47, 543–548.

Langkilde, A.M., Andersson, H., Bosaeus, I., 1993. Sugar-beet fibre increases cholesterol and reduces bile acid excretion from the small bowel. Br. J. Nutr. 70, 757–766.

Nordic Sugar, 2014. Sugar Beet Fibre Physiological Effects & Clinical Studies. Nordic Sugar AB, Malmo.

Nyman, M., Asp, N.G., 1982. Fermentation of dietary fibre components in the rat intestinal tract. Br. J. Nutr. 47, 357–366.

Overton, P.D., Furlonger, N., Beety, J.M., Chakraborty, J., Tredger, J.A., Morgan, L.M., 1994. The effects of dietary sugar-beet fibre and guar gum on lipid metabolism in Wistar rats. Br. J. Nutr. 72, 385–395.

Regulation (EC) No. 1924/2006, 2006. Nutrition and health claims made on foods, Regulation (EC) No. 1924/2006 of 20 December 2006 of the European Parliament and Council. Off. J. Eur. Union L12/3, 16.

Regulation (EU) No. 1169/2011, 2011. The provision of food information to consumers, Regulation (EU) No. 1169/2011 of 25 October 2011 of the European Parliament and Council amending Regulations (EC) No. 1924/2006 and (EC) No. 1925/2006 of the European Parliament and of the Council, and repealing Commission Directive 87/250/EEC, Council Directive 90/496/EEC, Commission Directive 1999/10/EC, Directive 2000/13/EC of the European Parliament and of the Council, Commission Directives 2002/67/EC and 2008/5/EC and Commission Regulation (EC) No. 608/2004. Off. J. Eur. Union L304, 18.

Rumney, C.J., Henderson, C., Stewart, C.S., 1990. Fermentation of wheat bran or sugar-beet fibre by human colonic bacteria growing in vitro in semi-continuous culture. Proc. Nut. Soc. 49, 124A.

SACN, 2008. Draft SACN Statement on Dietary Fibre and Health and the Dietary Fibre Definition. Department of Health, London, Available from: https://www.gov.uk/government/uploads/system/uploads/attachment_data/file/339367/SACN_Draft_position_statement_on_dietary_fibre_and_health_and_dietary_fibre_definition_332008.pdf.

SACN, 2015. Carbohydrates and Health. The Stationery Office, London, Available from: https://www.gov.uk/government/uploads/system/uploads/attachment_data/file/445503/SACN_Carbohydrates_and_Health.pdf.

Southgate, D., 1990. Dietary fibre and health. In: Southgate, Waldron, K., Johnson, I.T., Fenwick, G.R. (Eds.), Dietary Fibre: Chemical and Biological Aspects. Elsevier Science and Technology, Cambridge, UK, pp. 10–19, (Special Publication No. 83, Royal Society of Chemistry).

Stanner, S., 2015. 30 g of fibre a day: an achievable recommendation? Nutr. Bull. 40, 118–129.

Thibault, J.F., Renard, C.M.G.C., Guillion, F., 2001. Sugar beet fibre. In: Cho, S., Dreher, M.L. (Eds.), Handbook of Dietary Fiber. CRC Press, Boco Raton, FL, pp. 553–581.

Travis, J.S., Morgan, L.M., Tredger, J.A., Marks, V., 1990. Effects of sugar beet fibre on blood glucose serum lipids and apolipoproteins in non-insulin dependent diabetes mellitus. In: Southgate, D.A.T., Waldron, K., Johnson, I.T., Fenwick, G.R. (Eds.), Dietary Fibre: Chemical and Biological Aspects. Elsevier Science and Technology, Cambridge, UK, pp. 368–372, (Special Publication No. 83, Royal Society of Chemistry).

Tredger, J.A., Morgan, L.M., Travis, J., Marks, V., 1991. The effects of guar gum, sugar beet fibre and wheat bran supplementation on serum lipoprotein levels in normocholesterolaemic volunteers. J. Hum. Nutr. Diet. 4, 375–384.

Tredger, J., Sheard, C., Marks, V., 1981. Blood glucose and insulin levels in normal subjects following a meal with and without added sugar beet pulp. Diabetes Metab. 7, 169–172.

Winner, C., 1993. 1. History of the crop. In: Cooke, D.A., Scott, R.K. (Eds.), The Sugar Beet Crop. Chapman and Hall, London, pp. 1–32.

Further Reading

EC, 2006. In: Regulation (EC) No. 1924/2006 of the European Parliament and of the Council of 20 December 2006 on nutrition and health claims made on foods. Off. J. Eur. Union L12/3, 16.

US Food and Drug Administration, 2012. Sec. 101.77 Health claims: fruits, vegetables, and grain products that contain fiber, particularly soluble fiber, and risk of coronary heart disease. Title 21, Food and Drugs; Chapter I, Food and Drug Administration; Subchapter B, Food For Human Consumption; Part 101, Food Labeling; Subpart E, Specific Requirements for Health Claims. Code of Federal Regulations. US Government Publishing Office, Washington, DC. Available from: https://www.gpo.gov/fdsys/granule/CFR-2012-title2021-vol2012/CFR-2012-title2021-vol2012-sec2101-2077.

Authorised EU health claim for glycaemic carbohydrates and muscular recovery

9

Eric Chappuis
Naturalpha, Loos, France

9.1 Introduction: nutrition optimisation in physical activity at the crossroads of disciplines

Physical activity is defined as any bodily movement produced by skeletal muscles that requires energy expenditure (World Health Organization, 2016).

Participation in regular physical activity reduces the risk of coronary heart disease and stroke, diabetes, hypertension, colon cancer, breast cancer and depression. Additionally, physical activity is a key determinant of energy expenditure, and thus is fundamental to energy balance and weight control (World Health Organization, 2010).

Forty one percent of Europeans exercise or play sport at least once a week. The most common reason for engaging in sport or physical activity is to improve health (62%) but improving performance is also a strong motivation for 24% of EU citizens, especially in Northern European countries (European Commission, 2014).

Nutrition is a major component of physical performance and while many recommendations emphasise that a balanced diet in macro- and micronutrients can support physical performance through providing adequate energy, regulating properly metabolic functions and maintaining optimal body mass, scientific evidence suggests that particular foods or food ingredients beyond the general recommended dietary guidelines can improve these processes (Saris et al., 2003).

Improving physical performance or recovery can be envisaged through a wide array of benefits. The European Commission project Process for the Assessment of Scientific Support for Claims on Foods (PASSCLAIM) aimed at reviewing existing methods to support health claims in the European Union and identified several key areas to cover for foods and ingredients manufacturers: strength and power, endurance, hydration, flexibility tissue growth, immune function, energy supply and recovery (Saris et al., 2003).

9.2 Carbohydrates as key nutrients in physical activity

The role of carbohydrates in physical activity has been studied since the late 1960s as ergogenic aids in physical activity contexts (Bergstrom et al., 1967; Bergstrom and Hultman, 1967), as carbohydrates are (with fat) the primary source of energy for skeletal muscle during exercise (Cermak and van Loon, 2013).

Foods, Nutrients and Food Ingredients with Authorised EU Health Claims. http://dx.doi.org/10.1016/B978-0-08-100922-2.00009-7
Copyright © 2018 Elsevier Ltd. All rights reserved.

Carbohydrates include a range of components whose classification is generally based on molecular size and monomeric sugars with 3 main groups: sugars (1–2 monomers), oligosaccharides (3–9 monomers) and polysaccharides (10 monomers and more) (EFSA Panel on Dietetic Products Nutrition and Allergies, 2010).

9.2.1 Bioavailability of carbohydrates

Beyond the chemical classification of carbohydrates it is important to consider carbohydrates also with regards to their nutritional impact. Glycaemic carbohydrates are digested and absorbed in the small intestine, providing energy to body cells while non-digestible carbohydrates (also referred as to dietary fibres) pass into the large intestine to form substrate for the colonic microflora (EFSA Panel on Dietetic Products Nutrition and Allergies, 2010).

9.2.2 Role of carbohydrates in physical activity

Carbohydrates have legitimately received attention in sports nutrition as they display unique features in the performance of and adaptation to training.

Carbohydrates are a key substrate for the brain and central nervous system and a versatile substrate for muscular work. In this latter case, through their use by oxidative and anaerobic pathways carbohydrates support exercise over a wide range of intensities. Carbohydrates are the preferential energy source for exercise as they provide a higher yield of ATP compared with fat, leading to improved gross exercise efficiency.

Research data support the fact that the maintenance of high-carbohydrate availability during exercise relates to glycogen stores and blood glucose that improve performance of prolonged, sustained or intermittent high-intensity exercise. However, this support is limited by the availability of these carbohydrates, and the depletion of carbohydrate stores is associated with fatigue as witnessed by increased perception of effort, reduced work rates and impaired skill and concentration (Rodriguez et al., 2009).

Nutrition strategies to improve sports performance aim at ensuring an adequate supply of carbohydrates before and during exercise as well as during the recovery period to limit the depletion of carbohydrate stores (Rodriguez et al., 2009).

9.3 Measuring recovery after physical activity

Fatigue is defined as the loss of peak force or power output during an exercise. In this perspective, recovery is interpreted as the rate at which maximal force/power output is restored to normal after prior fatigue-inducing exercise (Saris et al., 2003).

In the context of the EU Regulation on Nutrition and Health Claims made on Foods (EC/1924/2006), the European Food Safety Authority (EFSA) recognises that the improvement, maintenance or reduced loss of muscle function (e.g. muscle strength) is a beneficial physiological effect. Applied to a recovery context, EFSA states that 'the faster recovery from muscle fatigue, muscle soreness or muscle damage after exercise contributing to the restoration of muscle function (e.g. muscle strength), can be considered beneficial physiological effects' (EFSA Panel on Dietetic Products Nutrition and Allergies, 2012).

9.3.1 Scientific assessment of recovery after physical activity

Evaluation of the effect of an intervention on recovery after physical activity consists of the evaluation of force/power output after an intermittent exercise protocol aimed at depressing the parameter investigated.

Several protocols have been developed to evaluate strength, power or anaerobic power and capacity. The most reliable method known to date is the evaluation of arm flexion or knee extension on an isokinetic dynamometer. It has a high reproducibility (coefficient of variation is only 1%–3%) and excellent precision, on the condition that the position of the arm or leg is standardised and the dynamometer used is reliable. Furthermore, the learning effect is negligible, it is independent of training status or age and no specific skills are required. Other methods include jump tests, sprint tests or other exercises on an isokinetic dynamometer but these are generally both less reproducible and precise (Saris et al., 2003).

9.3.2 EFSA health claims guidelines for recovery after physical activity

EFSA has released guidance on the scientific requirements for health claims related to physical performance (EFSA Panel on Dietetic Products Nutrition and Allergies, 2012). However, this guidance remains elusive on adequate protocols required to demonstrate an effect on muscular function.

EFSA proposes different combinations of types of evidence that can be provided to decipher the effect of an intervention on muscular function. Different exercises such as one repetition maximum weight lifting, isokinetic knee-extension torque or isometric handgrip strength could be accepted as an appropriate outcome measure of muscle function. Additionally changes in muscle structure (muscle mass, muscle shape, number and type of muscle fibres, muscle damage and muscle tissue repair) can be considered as beneficial effects, provided it is demonstrated that they contribute to the changes in muscle function claimed in the health claim application. Additionally, subjective measures of muscle fatigue or muscle soreness that are generally collected through specific questionnaires may be used as supportive evidence in this context.

9.4 Glycaemic carbohydrates and recovery of normal muscle function (contraction) after strenuous exercise

The particulars of the health claim authorised for use in the European Union, linking the intake of glycaemic carbohydrates and recovery of normal muscle function after strenuous exercise, are that they are based on an indirect evaluation of the benefit of the substance on the outcome. Indeed most of the claims aim at demonstrating a direct effect of a substance on the outcome which is the subject of the claim. In the specific case of this health claim, an indirect demonstration of the benefit was proposed through the causal link between the impact of glycogen status on exercise performance on one hand, and between the intake of glycaemic carbohydrates and their impact on glycogen stores on the other hand.

9.4.1 Assessing glycogen stores in skeletal muscles

Glycogen stores in skeletal muscles can be evaluated with different approaches. The first approach consists of calculating the net storage rate as an indicator of the resynthesis dynamics during a defined period of time. Muscle glycogen synthesis rate is calculated as the difference between glycogen concentrations post- and pre-recovery divided by the time of the recovery period (Roy and Tarnopolsky, 1998; Wallis et al., 2008). The second approach compares the glycogen concentration after recovery rather than the net storage rate as a marker of glycogen repletion.

Glycogen content in skeletal muscles can be evaluated by needle muscle biopsies and subsequent glycogen levels measurement (Adamo and Graham, 1998) or by ^{13}C nuclear magnetic resonance spectroscopy (^{13}C NMR or ^{13}C MRS) (Price et al., 1999).

9.4.2 Glycogen status influences exercise performance: clinical data

Glucose, which is mainly provided to body cells from glycaemic carbohydrates, can be stored as glycogen in the liver and in the muscles. This storage capacity is limited, in total to around 500 g for a typical adult person, of which 300–400 g can be stored in the muscles. While liver glycogen is essential for liver functions, and is used to maintain normal blood glucose concentrations between meals, muscle glycogen is used primarily as a source of energy within the muscles (EFSA Panel on Dietetic Products Nutrition and Allergies, 2010).

The relationship between nutritional supplementation and glycogen metabolism in an exercise situation has been investigated for decades. Early studies like that of Bergstrom et al. (1967) showed that muscle glycogen availability is an important factor determining the ability to perform long-term exercise, and that glycogen stores can substantially be modified by different diets (Bergstrom et al., 1967). As exercise intensity increases, the oxidation of carbohydrates becomes more predominant as a source of energy (Romijn et al., 1993), and the intensity of exercise cannot be maintained if glycogen stores are depleted (Ivy and Kuo, 1998). At an intensity of 60%–75% VO_2max, fatigue seems to be related to a combination of muscle glycogen depletion and hypoglycaemia, while at 75%–85% VO_2max, fatigue is mostly the result of glycogen depletion (Ivy, 1999). Thus, it appears that restoration of glycogen stores after exercise is an important factor in determining time of recovery (Jentjens and Jeukendrup, 2003).

9.4.3 Glycogen status influences exercise performance: mechanisms of action

The proper functioning of skeletal muscle function/contraction, and more precisely excitation–contraction (E–C) mechanisms, are highly dependent on muscular glycogen. This occurs through:

- action potential due to Na^+-K^+-ATPase activity
- sarcoplasmic reticulum (SR) calcium release
- levels of intracellular tetanic free calcium
- force production in muscles.

9.4.3.1 Effects on Na^+-K^+-ATPase activity

The Na^+-K^+ pumps in the transverse tubular system of a muscle fibre play a vital role in keeping K^+ concentration sufficiently low during activity to prevent depolarisation and induced loss of contractibility. These pumps are located in a key transduction zone (triad junction) that controls E–C coupling in the skeletal muscle. Dutka and Lamb (2007) showed in single mechanically skinned muscle fibres from rats that the functioning of these pumps was increased when phospho(enol)pyruvate (intermediate in glycolysis pathway that allows the transformation of ADP to ATP in cells catalysed by pyruvate kinase) was added to the medium, whatever the level of creatine kinase. This demonstrates on one hand that glycolytic resynthesis of ATP (which is linked to the availability of glycogen in muscle cells) is critical to meet the high demand of the Na^+-K^+ pump and maintain muscle excitability, and on the other hand the importance of the glycolysis pathway (which is dependent on glycogen levels in cells) to maintain low K^+ concentrations to limit loss of contractibility in muscle cells.

Glycogen and glycogen-derived molecules within muscular cells exert an important role in the regulation of Na^+-K^+-ATPase activity which is critical for the maintenance of correct cellular conditions for muscle contractions.

9.4.3.2 Sarcoplasmic reticulum (SR) calcium release

SR is an intracellular organelle and its main function is the control of muscle contraction through the regulation of intracellular calcium concentration. Calcium released from the SR to the cytosol induces contraction by binding to the actin filaments of muscle fibres. Following contraction, the calcium is then segregated into the SR through an energy-dependent Ca^{2+}-ATPase. This process of Ca^{2+} release and Ca^{2+} uptake, called Ca^{2+} cycling (Duhamel et al., 2006a), is impaired after strenuous exercise (Ortenblad et al., 2000). Intracellular Ca^{2+} handling is a major contributor to muscular loss of contractibility (Baker et al., 1993; Westerblad and Allen, 1996).

Experiments in humans following strenuous exercise (with depleted glycogen stores) have shown that:

- SR Ca^{2+} uptake in muscles after exercise was higher in subjects with higher muscular glycogen levels post-exercise compared with those with lower muscular glycogen levels (Duhamel et al., 2006b,c).

- The maximal muscular Ca^{2+}-ATPase activity is less decreased in subjects with high muscle glycogen content compared with those with lower muscular glycogen levels (Duhamel et al., 2006b,c).
- SR Ca^{2+} release rate is re-established in subjects with higher muscle glycogen levels 4 h post-exercise, while this does not occur in people with lower muscular glycogen levels (Duhamel et al., 2006c; Ortenblad et al., 2000).

These results show that glycogen levels in muscles are related to a limitation in the impairment of Ca^{2+} cycling after muscles have been solicited during strenuous exercise.

9.4.3.3 Levels of intracellular tetanic free calcium

In addition to SR Ca^{2+} release, intracellular free calcium concentration is involved in muscle fatigue. Several in vitro experiments have shown that the reductions in force, Ca^{2+} release and contractile protein inhibition observed during muscle fatigue are closely associated with reduced muscle glycogen concentration. In mouse muscle fibres submitted to glycogen-reducing contractions, decrease in tetanic force and tetanic calcium levels are only partially reversed in glucose-limiting conditions (associated with decreased glycogen levels), while this effect is delayed when glucose is added to the medium (Chin and Allen, 1997; Helander et al., 2002). Furthermore, Ca^{2+} sensitivity and maximum Ca^{2+}-activated force (Fmax) alterations due to stimulation are fully reversed when muscle glycogen recovers (Chin and Allen, 1997). In toad skeletal muscle, multiple fatigue runs in the absence of glucose were associated with a faster decline of the tetanic [Ca^{2+}] compared with a single fatigue run, suggesting an association between depletion of glycogen and accelerated decline of tetanic [Ca^{2+}]. Moreover, glucose addition enabled reversal of the faster fall of force in partly glycogen-depleted fibres (Kabbara et al., 2000).

When taken together these findings show that during recovery after muscular glycogen-reducing contractions, glycogen does not recover without glucose in the medium, and fibre bundles show reduced tetanic Ca^{2+} and endurance on further effort.

9.4.3.4 Force production in muscles

Muscular glycogen levels play a critical role in sustaining proper muscle contraction after muscle glycogen depletion, as evidenced by the results of various preclinical and clinical investigations.

In mechanically skinned rat muscle fibres, the intramyofibrillar glycogen levels were correlated to resistance exercise capacity (Stephenson et al., 1999), while some other experiments in humans have also shown an impaired capacity to perform exercise in glycogen-reducing conditions compared with glycogen restoration diets (Bangsbo et al., 1992; Bergstrom et al., 1967; Pernow and Saltin, 1971). Deeper mechanistic insights have been gained from use of different models. In toad muscle fibres, depletion of glycogen was correlated with a loss of the ability to respond to transversal tubular system depolarisation, even with high ATP and creatine phosphate levels in the medium. Consistently, initial glycogen levels are correlated to capacity to respond to such depolarisation (Stephenson et al., 1999). Intramyofibrillar glycogen levels have also been correlated to the resistance capacity of rat muscle fibres to physiological fatigue (Nielsen et al., 2009).

Altogether these data demonstrate the robust mechanistic link existing between muscle glycogen stores and muscular function.

9.4.4 Carbohydrate ingestion influences glycogen stores in the context of exercise performance: clinical data

A large number of intervention trials have assessed the influence of glycaemic carbohydrate ingestion on muscle glycogen recovery after a protocol of glycogen depletion induced by exercise. A total of 16 clinical trials (including 10 randomised clinical trials and 6 non-randomised controlled trials) have been retrieved, and are detailed in Table 9.1. The overall results are summarised here through the filters of:

- training level of individuals
- type of exercise considered
- timing of ingestion and duration of the effect
- dose ingested and magnitude of the effect
- types of glycaemic carbohydrates provided.

Considering the influence of training level as assessed by VO_2max, the available evidence does not suggest it influences the dynamics of muscular glycogen repletion after exercise (Bowtell et al., 2000; Hickner et al., 1997; Pascoe et al., 1993; van Hall et al., 2000; Wallis et al., 2008).

The type of exercise (and more precisely the type of muscular contraction) also does not seem to influence early muscle glycogen resynthesis rate after exercise (Jentjens and Jeukendrup, 2003) despite the lack of direct comparison within trials. Doyle et al. (1993) further investigated this topic by submitting athletes to exercise involving more eccentric or concentric muscle contraction, and assessing the glycogen repletion post-exercise. Surprisingly, they found that eccentric contractions lowered muscle glycogen resynthesis rate, but only 48 h after the end of exercise. It was suggested that muscle damage provoked by eccentric contraction may lower insulin action and glucose transport usually stimulated by muscle contraction (Doyle et al., 1993).

Considering the influence of the timing of ingestion, there is evidence that giving a first bolus immediately after exercise optimises the early muscle glycogen resynthesis rate (Ivy et al., 1988). Following exercise, it may be important to administer glycaemic carbohydrates regularly to keep plasma glucose and insulin concentrations elevated, which contributes to increase the muscle glycogen storage rate (van Loon et al., 2000). Providing subsequent glycaemic carbohydrate feeding every 2 h may not be sufficient to keep glucose and insulin concentrations elevated. Thus, shorter feeding times may be more adequate for maintaining a high synthesis rate (Beelen et al., 2010; Ivy et al., 1988; van Loon et al., 2000). Furthermore, glycogen replenishment to baseline levels cannot be achieved during a recovery period of less than 2 h, despite provision of exogenous glucose (Gusba et al., 2008; Haub et al., 1999;

Table 9.1 **Summary of interventional trials assessing the influence of glycaemic carbohydrates on skeletal muscle glycogen stores after strenuous exercise**

References	Subjects	VO_2max/1 RM	Ingestion timings	Total loads	Measures	Muscle glycogen	Synthesis rates	Muscle glycogen values	Synthesis rate values
Non-placebo controlled trials									
Haub et al. (1999)	6 Physically active males, cycling	50.8 ± 6.0 mL/kg/min	0 h	Maltdx versus plc 0.7 g/kg (51 g)	Biopsies 0, 1 h	No significant differences between treatments ($P > 0.05$)	NM	110–120 mmol/kg (w/w) at 60 min in both groups	NA
Gusba et al. (2008)	10 Physically active males, cycling	55 mL/kg/min	0, 60, 120, 180, 240 min	Gatorlode versus water 5 × 1 g/kg (400 g)	Biopsies 0, 30, 60, 120, 300 min	Significant higher in CHO at 120 min and 300 min ($P < 0.05$)	NM	266 mmol/kg d.w. at 300 min	NA
Rotman et al. (2000)	8 Moderately trained males, cycling	50.3 ± 8.7 mL/kg/min	0, 2 h	Glu:mix (suc, malt, glu, fru) versus water 1.7 g/kg (117 g)	^{13}C MRS 0, 4 h	NM	Significantly higher over 0–4 h in CHO versus placebo ($P = 0.014$)	NA	0.014 mUI/min
Ivy et al. (1988)	8 Healthy males, cycling	52 mL/kg/min	0, 2 h	Glu 2 × 1.5 g/kg (226 g) 2 × 3 g/kg (452 g) 2 × 0 g/kg (plc)	Biopsies 0, 2, 4 h	NM	Significantly higher glycogen storage rate in both CHO groups compared with placebo, during the first and second 2 -h recovery ($P < 0.05$); no significant differences between	NA	During the first 2 h [in mmol/kg (w/w)] 5.2 ± 0.9 in 226 g 5.8 ± 0.7 in 452 g

Pascoe et al. (1993)	8 Recreationnally active males, resistance exercise	48 ± 2.6 kg	0, 1 h	CHO versus water 2 × 1.5 g/kg (240 g)	Biopsies 0, 2, 6 h	Significantly higher in CHO at 6 h ($P < 0.05$), not at 2 h ($P > 0.05$)	Higher during the first 2 h in CHO ($P < 0.05$)	In mmol/kg (w/w) After 2 h CHO: 117.6 ± 16.5 After 2 h plc: 105.1 ± 13.1 After 6 h CHO: 123.4 ± 15.1 After 2 h plc: 105.5 ± 13.0	During the first 2 h [in mmol/kg (w/w)/h] 12.9 ± 4 in CHO 1.9 ± 2.2 in plc
Roy and Tarnopolsky (1998)	10 Healthy males, resistance exercise	NM	0, 1 h	Suc:glu (56%:44%) versus plc 2 × 1 g/kg (174 g)	Biopsies 0, 4 h	Non-significantly increased at 4 h in CHO compared with placebo ($P > 0.05$)	Significantly higher over 0–4 h in CHO versus placebo ($P < 0.05$)	Appr 310 mmol/kg d.m. versus 260 mmol/kg d.m.	In mmol/kg d.w./h 20 versus 2
van Hall et al. (2000)	5 Healthy males, cycling	61 ± 2 mL/kg/min	Every 15 min	Suc versus water 4.8 g/kg (350 g)	Biopsies 0, 1.5, 4 h	Higher at 1.5 and 4 h in CHO compared with water ($P > 0.05$)	Significantly higher at 1.5 and 4 h in CHO versus water ($P > 0.05$)	In mmol/kg d.w. After 1.5 h CHO: 170 After 1.5 h plc: 100 After 4 h CHO: 250 After 4 h plc: 120	In mmol/kg d.w./h For 0–1.5 h CHO: 50 For 0–1.5 h plc: 20 For 1.5–4 h CHO: 40 For 0–1.5 hplc: 10

(Continued)

Table 9.1 Summary of interventional trials assessing the influence of glycaemic carbohydrates on skeletal muscle glycogen stores after strenuous exercise (*cont.*)

References	Subjects	VO_2max/1 RM	Ingestion timings	Total loads	Measures	Muscle glycogen	Synthesis rates	Muscle glycogen values	Synthesis rate values
Casey et al. (2000)	10 Well-trained males cycling	53 mL/kg/min	0 h	Suc, gluc or placebo 1 g/kg (76 g)	^{13}C MRS Hourly	No significant differences between conditions for both muscle and liver glycogen ($P > 0.05$)	Significantly higher ($P \leq 0.01$) only during the first hour in the sucrose trial compared with placebo for liver glycogen only	In mmol/L at 4 h 92 ± 38 in glu 80 ± 10 in suc 68 ± 8 in plc	In mmol/L/h at 1 h 40 ± 39 in glu 5 ± 9 in suc −12 ± 21 in plc In mmol/L/h at 4 h −1 ± 4 in glu 8 ± 9 in suc 4 ± 11 in plc
Tarnopolsky et al. (1997)	8 Males and females, cycling	Males: 56.9 ± 2.65 Females: 51.7 ± 2.4 mL/kg/min	0, 1 h	Suc/glu versus placebo 2 × 1 g/kg (146 g for males and 122 g females)	Biopsies 0, 4 h	NM	Significantly higher in the CHO group compared with the placebo for both genders ($P < 0.01$)	NA	In mmol/kg d.m./h 40.0 in CHO men 34.5 in CHO women 3.0 in plc men 12.0 in plc women

Bowtell et al. (2000)	7 Untrained males cycling	42.3 ± 1.5 mL/min/kg	0 h	0.8 g/kg glu (62 g) 0.8 g/kg suc (62 g) 0.5 g/kg suc (38.5 g)	Biopsies and ^{13}C label-ling 0, 1, 2 h	NA (appears to be higher in glu at 2 h)	Significantly higher in the glucose trial over recovery ($P < 0.05$)	Appr 22 mmol/kg (w/w) inglu versus 12 and 16 mmol/kg (w/w) in LS and HS	In mmol/kg (w/w)/h 5.31 ± 1.11 in glu 4.07 ± 1.05 in HS 3.45 ± 0.85 in LS
Howarth et al. (2009)	6 Healthy males, cycling	49 mL/min/kg	15 min intervals for 3 h	1.2 g/kg/h maltdx (324 g) 1.6 g/kg/h maltdx (432 g)	Biopsies 0, 4 h	No differences between trials ($P > 0.05$), both significantly increased compared with baseline	No differences between trials ($P > 0.05$)	Appr 180 mmol/kg d.w. in both trials	In mmol/kg d.w./h 23 ± 3 in 324 g 25 ± 7 in 432 g
Blom et al. (1987)	27 Healthy males, cycling	56 mL/kg/min	0, 2, 4 h	3 × 0.35 g/kg, glu (LG– 79 g) 3 × 0.7 g/kg, glu (MG–157 g) 3 × 0.7 g/kg, suc (MS–157 g) 3 × 0.7 g/kg, fru (MF–157 g) 3 × 1.4 g/kg, glu (HG–315 g)	Biopsies 0, 2, 4, 6 h	NM	LG × 2 = MG = HG ($P > 0.05$) MF < MG ($P < 0.05$) MG = MS ($P > 0.05$)	NA	In mmol/kg (w/w)/h MG: 5.8 ± 1.0 HG: 5.7 ± 0.9 MS: 6.2 ± 0.5 MF: 3.2 ± 0.7

(*Continued*)

Table 9.1 **Summary of interventional trials assessing the influence of glycaemic carbohydrates on skeletal muscle glycogen stores after strenuous exercise (*cont.*)**

References	Subjects	VO_2max/1 RM	Ingestion timings	Total loads	Measures	Muscle glycogen	Synthesis rates	Muscle glycogen values	Synthesis rate values
Wallis et al. (2008)	6 Healthy males engaged in endurance training cycling	68 mL/min/kg	Every 30 min for 4 h	1.2 g/kg/h, glu (G–360 g) 1.2 g/kg/h, glu + fru 2:1 (GF–360 g)	Biopsies 0, 1, 4 h	No differences between trials at 1 h and 4 h ($P > 0.05$) Trials significantly higher compared with baseline at 4 h only ($P < 0.05$)	No differences between trials over the 4 h ($P > 0.05$)	In mmol/kg d.m. 303 ± 46 in G 267 ± 38 in GF	In mmol/kg d.m./h 44 ± 8 in G 39 ± 8 in GF
van Loon et al. (2000)	8 Well-trained males, cycling	390 ± 8 W (maximal workload)	Every 30 min during 270 min	0.8 g/kg/h glu/maltdx (1:1) 280 g 1.2 g/kg/h glu/maltdx (1:1) 420 g	Biopsies 0, 5 h	Higher in the 1.2 g/kg trial compared with the 0.8 g/kg at 5 h ($P < 0.05$)	Higher in the 1.2 g/kg trial compared with the 0.8 g/kg over 5 h ($P < 0.05$)	In mmol/kg d.m. at 5 h 362 ± 46 for 420 g 272 ± 54 for 280 g	In mmol/kg d.m./h 44.8 ± 6.8 for 420 g 16.6 ± 7.8 for 280 g

Tsintzas et al. (2003)	7 Recreational males, running	56 ± 2 mL/kg/min	0, 60, 90, 120, 150, 180 min	Solution (dxt, fru, malt, saccharides and elec) 2.2 g/kg (175 g) high 0.6 g/kg (50 g) low	Biopsies 0, 4 h	Higher muscle glycogen content at 4 h in 175 g compared with 50 g ($P < 0.01$)	Higher muscle glycogen synthesis at 4 h in 175 g compared with 50 g ($P < 0.01$)	In mmol/kg d.m. 334 ± 43 in high 283 ± 44 in low	In mmol/kg d.m./h 75 ± 20 in high 31 ± 11 in low
Bilzon et al. (2002)	11 Trained males, running	60.7 ± 1.8 mL/min/kg	0, 60, 120, 180 min	0.7 g/kg glu (55 g) 3 g/kg glu (220 g)	^{13}C labelling 0, 4 h	NM	Fivefold greater in the 220 g trial compared with the 55 g trial ($P < 0.01$)	NA	Glycogen synthesised over the 4 h 176 ± 3.6 g in 220 g 33 ± 2.7 g in 55 g

Appr, Approximately; CHO, carbohydrate; dxt, dextran; d.m., dry matter; d.w., dry weight; elec, electrolytes; glu, glucose; GF, glucose-fructose; fru, fructose; HF, high fructose; HG, high glucose; HS, high sucrose; malt, maltose; MF, medium fructose; MG, medium glucose; MRS, magnetic resonance spectroscopy; NA, not applicable; NM, not mentioned; plc, placebo; RM, repetition max; suc, sucrose; w/w, weight by weight

Pascoe et al., 1993; Wallis et al., 2008) and should last at least 4 h to be optimised (Rotman et al., 2000; Roy and Tarnopolsky, 1998; Tarnopolsky et al., 1997; van Hall et al., 2000). While the optimal feeding frequency remains to be determined, ingesting glycaemic carbohydrates immediately after strenuous exercise and every hour thereafter seems to be adequate to maintain a high glycogen synthesis rate.

Considering the influence of the dose provided in the different trials reviewed, current evidence suggests that glycaemic carbohydrates given at doses from 100 to 220 g over a recovery period of 4–6 h significantly increase the glycogen repletion post-exercise (Casey et al., 2000; Ivy et al., 1988; Rotman et al., 2000; Roy and Tarnopolsky, 1998; Tarnopolsky et al., 1997; van Hall et al., 2000). Studies suggest an increased effect when giving a medium dose of glycaemic carbohydrates (100–220 g) compared with a low dose (<100 g) (Bilzon et al., 2002; Tsintzas et al., 2003). However, the effect tends to plateau when high and very high doses are given (Blom et al., 1987; Bowtell et al., 2000; Howarth et al., 2009; Ivy et al., 1988). A potential explanation is that carbohydrate availability, and more specifically intestinal glucose absorption, may be a rate-limiting factor for muscle glycogen synthesis (Jentjens and Jeukendrup, 2003).

Considering the influence of type of glycaemic carbohydrate, glucose and glucose polymers are effective for restoring glycogen stores shortly after exercise (Bilzon et al., 2002; Blom et al., 1987; Haub et al., 1999; Howarth et al., 2009; Ivy et al., 1988; van Loon et al., 2000; Wallis et al., 2008). There is evidence that fructose elicits a lower response on glycogen repletion than glucose or sucrose, though its effect remains significant (Blom et al., 1987). Generally, a combination of different glycaemic carbohydrates has been shown to be effective for glycogen repletion (Rotman et al., 2000; Tarnopolsky et al., 1997; van Loon et al., 2000; Wallis et al., 2008). Carbohydrates with high- and moderate-glycaemic index show slightly higher glycogen repletion compared with a low-glycaemic index diet (Blom et al., 1987; Bowtell et al., 2000) probably due to the higher rise in plasma glucose and insulin which exacerbates muscle membrane permeability. Overall, all glycaemic carbohydrates seem to exert beneficial effects on muscle glycogen replenishment after strenuous exercise and the evidence for a difference between carbohydrates is still limited.

Taken together, the results of these intervention trials demonstrate that the intake of glycaemic carbohydrates at doses from 1.7 to 4 g/kg of body weight given every hour for at least 4 h following strenuous exercise improves the replenishment of muscular glycogen stores, whatever the type of exercise considered, the training levels of individuals and the type of glycaemic carbohydrates.

9.4.5 Carbohydrate ingestion influences glycogen stores in the context of exercise performance: mechanisms of action

A large number of studies are consistent with the observation that muscle glycogen resynthesis after glycogen-depleting exercise follows a biphasic pattern (Jentjens and Jeukendrup, 2003).

From the end of exercise to 30–60 min later, rapid glycogen resynthesis occurs, probably reflecting increased glycogen synthase activity despite the absence of insulin. This phase refers to the insulin-independent phase (Jentjens and Jeukendrup, 2003).

Glycogen synthase activity can be stimulated by both muscle contraction and insulin, but it has been shown in several in vivo investigations that these stimulatory effects are under the influence of decreased glycogen content which is a potent regulator of glycogen synthase activity (Nielsen et al., 2009). Apart from an effect on glycogen synthase, the post-exercise state is characterised by a residual increase in glucose uptake through a contraction-induced GLUT-4 translocation in the skeletal muscles. The mechanisms remain unclear and involve a high number of molecular signals (Jentjens and Jeukendrup, 2003; Jensen and Richter, 2012).

The slow phase of muscle glycogen synthesis or insulin-dependent phase refers to the ability of muscle contraction to stimulate the sensitivity of muscle glucose uptake to insulin. This phenomenon could facilitate glucose uptake in the post-exercise state, when a meal is provided, allowing replenishment of glycogen stores. However, the mechanisms behind this effect are not fully understood, but could involve repackaging of the GLUT-4 vesicles (Jensen and Richter, 2012; Jensen et al., 2011). The slow phase of muscle glycogen resynthesis can last up to 48 h, while the rapid phase generally resolves within the first hour post-exercise.

The ingestion of glycaemic carbohydrates provokes an immediate increase in plasma insulin concentration (EFSA Panel on Dietetic Products Nutrition and Allergies, 2010). Ingestion of such carbohydrates immediately after exercise results in a rise in blood glucose which is concomitant to the increased plasma insulin. The effect of the latter on glycogen storage may be exacerbated by the combination of the increased insulin sensitivity and membrane permeability of muscle cells provoked by exercise. Simultaneously, the increased glycaemia allows the muscle cells to uptake glucose and replenish their glycogen stores (Richter et al., 1989).

9.4.6 EFSA and European Commission considerations on the health claim

Taken together the research data detailed previously demonstrate that glycaemic carbohydrates improve muscle glycogen repletion after strenuous exercise.

EFSA published a favourable opinion on the health claim application. It proposed conditions of use of doses of 4 g/kg of body weight in the first 4–6 h following strenuous exercise, with no restriction on the target population, provided subjects have performed strenuous exercise (EFSA Panel on Dietetic Products Nutrition and Allergies, 2013).

The European Commission followed EFSA's opinion and authorised the following wording and conditions of use:

- Wording: 'Carbohydrates contribute to the recovery of normal muscle function (contraction) after highly intensive and/or long-lasting physical exercise leading to muscle fatigue and the depletion of glycogen stores in skeletal muscle'.
- Conditions of use: The claim may be used only for food which provides carbohydrates which are metabolised by humans (excluding polyols). Information shall be given to the consumer that the beneficial effect is obtained with the consumption of carbohydrates, from all sources, at a total intake of 4 g/kg body weight, at doses, within the first 4 h and not later than 6 h, following a highly intensive and/or long-lasting physical exercise leading to muscle

fatigue and the depletion of glycogen stores in skeletal muscle. The claim may be used only for foods intended for adults who have performed highly intensive and/or long-lasting physical exercise leading to muscle fatigue and the depletion of glycogen stores in skeletal muscle.

References

Adamo, K.B., Graham, T.E., 1998. Comparison of traditional measurements with macroglycogen and proglycogen analysis of muscle glycogen. J. Appl. Physiol. 84 (3), 908–913.

Baker, A.J., Longuemare, M.C., Brandes, R., Weiner, M.W., 1993. Intracellular tetanic calcium signals are reduced in fatigue of whole skeletal muscle. Am. J. Physiol. 264 (3 Pt. 1), C577–C582.

Bangsbo, J., Norregaard, L., Thorsoe, F., 1992. The effect of carbohydrate diet on intermittent exercise performance. Int. J. Sports. Med. 13 (2), 152–157.

Beelen, M., Burke, L.M., Gibala, M.J., van Loon, L.J., 2010. Nutritional strategies to promote postexercise recovery. Int. J. Sport. Nutr. Exerc. Metab. 20 (6), 515–532.

Bergstrom, J., Hermansen, L., Hultman, E., Saltin, B., 1967. Diet, muscle glycogen and physical performance. Acta. Physiol. Scand. 71 (2), 140–150.

Bergstrom, J., Hultman, E., 1967. A study of the glycogen metabolism during exercise in man. Scand. J. Clin. Lab. Invest. 19 (3), 218–228.

Bilzon, J.L., Murphy, J.L., Allsopp, A.J., Wootton, S.A., Williams, C., 2002. Influence of glucose ingestion by humans during recovery from exercise on substrate utilisation during subsequent exercise in a warm environment. Eur. J. Appl. Physiol. 87 (4–5), 318–326.

Blom, P.C., Hostmark, A.T., Vaage, O., Kardel, K.R., Maehlum, S., 1987. Effect of different post-exercise sugar diets on the rate of muscle glycogen synthesis. Med. Sci. Sports Exerc. 19 (5), 491–496.

Bowtell, J.L., Gelly, K., Jackman, M.L., Patel, A., Simeoni, M., Rennie, M.J., 2000. Effect of different carbohydrate drinks on whole body carbohydrate storage after exhaustive exercise. J. Appl. Physiol. 88 (5), 1529–1536.

Casey, A., Mann, R., Banister, K., Fox, J., Morris, P.G., Macdonald, I.A., et al., 2000. Effect of carbohydrate ingestion on glycogen resynthesis in human liver and skeletal muscle, measured by ^{13}C MRS. Am. J. Physiol. Endocrinol. Metab. 278 (1), E65–E75.

Cermak, N.M., van Loon, L.J., 2013. The use of carbohydrates during exercise as an ergogenic aid. Sports Med. 43 (11), 1139–1155.

Chin, E.R., Allen, D.G., 1997. Effects of reduced muscle glycogen concentration on force, Ca^{2+} release and contractile protein function in intact mouse skeletal muscle. J. Physiol. 498 (Pt 1), 17–29.

Doyle, J.A., Sherman, W.M., Strauss, R.L., 1993. Effects of eccentric and concentric exercise on muscle glycogen replenishment. J. Appl. Physiol. 74 (4), 1848–1855.

Duhamel, T.A., Green, H.J., Perco, J.G., Ouyang, J., 2006a. Comparative effects of a low-carbohydrate diet and exercise plus a low-carbohydrate diet on muscle sarcoplasmic reticulum responses in males. Am. J. Physiol. Cell Physiol. 291 (4), C607–C617.

Duhamel, T.A., Green, H.J., Perco, J.G., Ouyang, J., 2006b. Effects of prior exercise and a low-carbohydrate diet on muscle sarcoplasmic reticulum function during cycling in women. J. Appl. Physiol. 101 (3), 695–706.

Duhamel, T.A., Perco, J.G., Green, H.J., 2006 Octc. Manipulation of dietary carbohydrates after prolonged effort modifies muscle sarcoplasmic reticulum responses in exercising males. Am. J. Physiol. Regul. Integr. Comp. Physiol. 291 (4), R1100–R1110, http://www.ncbi.nlm.nih.gov/entrez/query.fcgi?cmd=Retrieve&db=PubMed&dopt=Citation&list_uids=16690765.

Dutka, T.L., Lamb, G.D., 2007. Na^+-K^+ pumps in the transverse tubular system of skeletal muscle fibers preferentially use ATP from glycolysis. Am. J. Physiol. Cell Physiol. 293 (3), C967–C977.

EFSA Panel on Dietetic Products Nutrition and Allergies, 2010. Scientific Opinion on dietary reference values for carbohydrates and dietary fibre. EFSA J. 8 (3), 77.

EFSA Panel on Dietetic Products Nutrition and Allergies, 2012. Guidance on the scientific requirements for health claims related to physical performance. EFSA J. 10 (7), 9.

EFSA Panel on Dietetic Products Nutrition and Allergies, 2013. Scientific Opinion on the substantiation of a health claim related to glycaemic carbohydrates and recovery of normal muscle function (contraction) after strenuous exercise pursuant to Article 13(5) of Regulation (EC) No. 1924/2006. EFSA J. 11 (10), 3409–3419, http://dx.doi.org/10.2903/j.efsa.2013.3409.

European Commission, 2014. Special Eurobarometer 412 "Sport and physical activity".

Gusba, J.E., Wilson, R.J., Robinson, D.L., Graham, T.E., 2008. Interleukin-6 and its mRNA responses in exercise and recovery: relationship to muscle glycogen. Scand. J. Med. Sci. Sports 18 (1), 77–85.

Haub, M.D., Potteiger, J.A., Jacobsen, D.J., Nau, K.L., Magee, L.A., Comeau, M.J., 1999. Glycogen replenishment and repeated maximal effort exercise: effect of liquid carbohydrate. Int. J. Sport Nutr. 9 (4), 406–415.

Helander, I., Westerblad, H., Katz, A., 2002. Effects of glucose on contractile function, $[Ca^{2+}]$ i, and glycogen in isolated mouse skeletal muscle. Am. J. Physiol. Cell Physiol. 282 (6), C1306–C1312.

Hickner, R.C., Fisher, J.S., Hansen, P.A., Racette, S.B., Mier, C.M., Turner, M.J., et al., 1997. Muscle glycogen accumulation after endurance exercise in trained and untrained individuals. J. Appl. Physiol. 83 (3), 897–903.

Howarth, K.R., Moreau, N.A., Phillips, S.M., Gibala, M.J., 2009. Coingestion of protein with carbohydrate during recovery from endurance exercise stimulates skeletal muscle protein synthesis in humans. J. Appl. Physiol. 106 (4), 1394–1402.

Ivy, J.L., 1999. Role of carbohydrate in physical activity. Clin. Sports Med. 18 (3), 469–484.

Ivy, J.L., Katz, A.L., Cutler, C.L., Sherman, W.M., Coyle, E.F., 1988. Muscle glycogen synthesis after exercise: effect of time of carbohydrate ingestion. J. Appl. Physiol. 64 (4), 1480–1485.

Ivy, J.L., Kuo, C.H., 1998. Regulation of GLUT4 protein and glycogen synthase during muscle glycogen synthesis after exercise. Acta. Physiol. Scand. 162 (3), 295–304.

Jentjens, R., Jeukendrup, A., 2003. Determinants of post-exercise glycogen synthesis during short-term recovery. Sports Med. 33 (2), 117–144.

Jensen, T.E., Richter, E.A., 2012. Regulation of glucose and glycogen metabolism during and after exercise. J. Physiol. 590 (Pt 5), 1069–1076.

Jensen, J., Rustad, P.I., Kolnes, A.J., Lai, Y.C., 2011. The role of skeletal muscle glycogen breakdown for regulation of insulin sensitivity by exercise. Front. Physiol. 2, 112.

Kabbara, A.A., Nguyen, L.T., Stephenson, G.M., Allen, D.G., 2000. Intracellular calcium during fatigue of cane toad skeletal muscle in the absence of glucose. J. Muscle Res. Cell Motil. 21 (5), 481–489.

Nielsen, J., Schroder, H.D., Rix, C.G., Ortenblad, N., 2009. Distinct effects of subcellular glycogen localization on tetanic relaxation time and endurance in mechanically skinned rat skeletal muscle fibres. J. Physiol. 587 (Pt 14), 3679–3690.

Ortenblad, N., Sjogaard, G., Madsen, K., 2000. Impaired sarcoplasmic reticulum Ca^{2+} release rate after fatiguing stimulation in rat skeletal muscle. J. Appl. Physiol. 89 (1), 210–217.

Pascoe, D.D., Costill, D.L., Fink, W.J., Robergs, R.A., Zachwieja, J.J., 1993. Glycogen resynthesis in skeletal muscle following resistive exercise. Med. Sci. Sports Exerc. 25 (3), 349–354.

Price, T.B., Rothman, D.L., Shulman, R.G., 1999. NMR of glycogen in exercise. Proc. Nutr. Soc. 58 (4), 851–859.

Pernow, B., Saltin, B., 1971. Availability of substrates and capacity for prolonged heavy exercise in man. J. Appl. Physiol. 31 (3), 416–422.

Richter, E.A., Mikines, K.J., Galbo, H., Kiens, B., 1989. Effect of exercise on insulin action in human skeletal muscle. J. Appl. Physiol. 66 (2), 876–885.

Rodriguez, N.R., Di Marco, N.M., Langley, S., 2009. American College of Sports Medicine position stand. Nutrition and athletic performance. Med. Sci. Sports Exerc. 41 (3), 709–731.

Romijn, J.A., Coyle, E.F., Sidossis, L.S., Gastaldelli, A., Horowitz, J.F., Endert, E., et al., 1993. Regulation of endogenous fat and carbohydrate metabolism in relation to exercise intensity and duration. Am. J. Physiol. 265 (3 Pt 1), E380–E391.

Rotman, S., Slotboom, J., Kreis, R., Boesch, C., Jequier, E., 2000. Muscle glycogen recovery after exercise measured by ^{13}C-magnetic resonance spectroscopy in humans: effect of nutritional solutions. MAGMA 11 (3), 114–121.

Roy, B.D., Tarnopolsky, M.A., 1998. Influence of differing macronutrient intakes on muscle glycogen resynthesis after resistance exercise. J. Appl. Physiol. 84 (3), 890–896.

Saris, W.H., Antoine, J.M., Brouns, F., Fogelholm, M., Gleeson, M., Hespel, P., et al., 2003. PASSCLAIM—physical performance and fitness. Eur. J. Nutr. 42 (Suppl. 1), I50–195.

Stephenson, D.G., Nguyen, L.T., Stephenson, G.M., 1999. Glycogen content and excitation-contraction coupling in mechanically skinned muscle fibres of the cane toad. J. Physiol. 519 (Pt 1), 177–187.

Tarnopolsky, M.A., Bosman, M., Macdonald, J.R., Vandeputte, D., Martin, J., Roy, B.D., 1997. Postexercise protein-carbohydrate and carbohydrate supplements increase muscle glycogen in men and women. J. Appl. Physiol. 83 (6), 1877–1883.

Tsintzas, K., Williams, C., Boobis, L., Symington, S., Moorehouse, J., Garcia-Roves, P., et al., 2003. Effect of carbohydrate feeding during recovery from prolonged running on muscle glycogen metabolism during subsequent exercise. Int. J. Sports Med. 24 (6), 452–458.

van Hall, G., Shirreffs, S.M., Calbet, J.A., 2000. Muscle glycogen resynthesis during recovery from cycle exercise: no effect of additional protein ingestion. J. Appl. Physiol. 88 (5), 1631–1636.

van Loon, L.J., Saris, W.H., Kruijshoop, M., Wagenmakers, A.J., 2000. Maximizing postexercise muscle glycogen synthesis: carbohydrate supplementation and the application of amino acid or protein hydrolysate mixtures. Am. J. Clin. Nutr. 72 (1), 106–111.

Wallis, G.A., Hulston, C.J., Mann, C.H., Roper, H.P., Tipton, K.D., Jeukendrup, A.E., 2008. Postexercise muscle glycogen synthesis with combined glucose and fructose ingestion. Med. Sci. Sports Exerc. 40 (10), 1789–1794.

Westerblad, H., Allen, D.G., 1996. Mechanisms underlying changes of tetanic $[Ca^{2+}]i$ and force in skeletal muscle. Acta. Physiol. Scand. 156 (3), 407–416.

World Health Organization, 2010. Global recommendations on physical activity for health.

World Health Organization. Physical activity, 2016. Available from: http://www.who.int/topics/physical_activity/en/.

Authorised EU health claim for chicory inulin

10

Stephan Theis
BENEO Institute, Obrigheim, Germany

10.1 Introduction

Dietary fibre has a major role in bowel function and low fibre intake has been linked to gastro-intestinal symptoms such as constipation (EFSA, 2010b; IoM, 2005). With a prevalence of 5%–30%, constipation is one of the most common health impairments in western countries and is associated with a low frequency of bowel movements (Candelli et al., 2001; Cummings et al., 2004). According to EFSA's Guidance documents, changes in bowel function such as more frequent bowel movements are thus considered beneficial physiological effects for the general population, provided they do not result in diarrhoea (EFSA, 2011).

Public awareness of the importance of dietary fibre intake and maintaining digestive health has been steadily growing in recent years. The vast majority of European consumers find a healthy digestive system an important topic according to recent quantitative consumer research (BENEO, 2016).

The health benefits of chicory-derived inulin-type fructans for digestive health have been the subject of research for more than 20 years. Inulin-type fructans are prebiotic dietary fibres, naturally occurring in many food plants. The chicory root with its high content of inulin-type fructans is the most common plant raw material for providing inulin. Inulin-type fructans are non-digestible in the small intestine and fermented in the large intestine, resulting in selective modifications in the colonic microbiota composition and in the production of several metabolites. Thereby, they are among the very few scientifically proven and established prebiotics. Accordingly, inulin is a dietary fibre with established prebiotic effects (Micka et al., 2016; Moser et al., 2015).

Positive effects of inulin on digestive health and bowel function have been reported by international research groups in several human intervention trials as well as mechanistically oriented studies. Recently these benefits were also confirmed in a meta-analysis of randomised controlled trials (Collado Yurrita et al., 2014). This meta-analysis of five human intervention studies found a significant overall effect of inulin on bowel motor function, including significant improvements in stool frequency, stool consistency, transit time and hardness of stool (Collado Yurrita et al., 2014). Data are also supportive of more regular and/or softer stools in infants and young children (Closa-Monasterolo et al., 2013, 2016; Moser et al., 2015). Beneficial effects of daily supplementation with Orafti Inulin-type fructans have also recently been confirmed in 2- to 5-year-old constipated children (Closa-Monasterolo et al., 2016).

Foods, Nutrients and Food Ingredients with Authorised EU Health Claims. http://dx.doi.org/10.1016/B978-0-08-100922-2.00010-3
Copyright © 2018 Elsevier Ltd. All rights reserved.

The beneficial physiological effect of chicory inulin on bowel function (increased stool frequency) in the general population has been the subject of an Article 13.5 health claim application by BENEO leading to a positive scientific opinion by EFSA Panel on Dietetic Products, Nutrition and Allergies (NDA) in 2015 (EFSA, 2015). Following scientific approval by EFSA, the EU Commission authorised a corresponding health claim for 'native chicory inulin' from 1 January 2016 with proprietary use for BENEO's Orafti Inulin for a period of 5 years. This has been the first authorised health claim for a prebiotic fibre and is one of only few examples of approved health claims for dietary fibre according to the EU Nutrition and Health Claims Regulation.

10.2 Authorised EU health claim

Following an application from BENEO submitted in May 2014 pursuant to Article 13.5 of Regulation (EC) No. 1924/2006 and including a request for the protection of proprietary data, EFSA published a positive scientific opinion on chicory inulin and bowel function (increased stool frequency) in January 2015 (EFSA, 2015). The EFSA NDA Panel concluded that on the basis of the data presented 'a cause and effect relationship has been established' and noted 'the plausible mechanisms by which inulin and inulin-type fructans in "native chicory inulin" could exert the claimed effect'. EFSA confirmed that the effect on bowel function demonstrated by the intake of native chicory inulin is a beneficial physiological effect, whilst no adverse effects were reported in any of the studies (EFSA, 2015).

On 12 December 2015 the Commission Regulation (EU) 2015/2314 authorising the Article 13.5 claim on chicory inulin and bowel function (increased stool frequency) was published. The specific health claim wording approved under Article 13.5 reads as follows: 'Chicory inulin contributes to normal bowel function by increasing stool frequency'. The Regulation granted the proprietary use of this claim to BENEO and its customers for a period of 5 years from 1 January 2016 onwards [Regulation (EU) 2015/2314]. The dosage accepted in the European Union specifically for the approved claim was 12 g/day. This was based on new data proprietary to BENEO from a randomised controlled human trial.

In addition to the authorised proprietary Article 13.5 health claim, general health-related, well-being claims according to Article 10.3 of Regulation (EC) 1924/2006 are also possible along with the specific health claim. Amongst others, these include 'chicory inulin promotes digestive health' and 'chicory root fibre supports a healthy and balanced digestive system'.

10.3 Proprietary data

The health claim application from BENEO pursuant to Article 13.5 of Regulation (EC) 1924/2006 contained a request for the protection of proprietary data for the unpublished study report by Schulz and Schön (2012) in accordance with Article 21.

According to the EFSA NDA Panel, this study was required to establish the conditions of use for this specific health claim (EFSA, 2015; Regulation (EU) 2015/2314).

The study by Schulz and Schön (2012) was a double-blind, randomised, placebo-controlled human intervention study undertaken on behalf of and funded by BENEO. BENEO is the sole owner of the corresponding results and rights in the use of this data. Following authorisation of the health claim and granting of protection of proprietary data status, the study results have been submitted to and were published, following a peer review by the International Journal of Food Sciences and Nutrition (Micka et al., 2016).

The study in mildly constipated subjects was specifically designed to serve as a key human intervention study for the health claim application. It followed recent guidance documents for appropriate study designs to investigate digestive function (EFSA, 2011; US Department of Health and Human Services Food and Drug Administration, 2012). The study was registered at clinicaltrials.gov, a service of the US National Institutes of Health, as NCT02548247.

The main objective of this trial was to determine the effect of a 4-week daily consumption of Orafti Inulin from chicory on the frequency of bowel movements. Moreover, the effects of consumption of Orafti Inulin on stool consistency, gastrointestinal characteristics and health-related quality of life were assessed as secondary outcome parameters and were compared with the consumption of maltodextrin, a fully digestible carbohydrate used as the placebo control product (reference) for which no effects on bowel motor function are known. To minimise inter-individual differences, the trial was performed as a two-way cross-over design.

Each of the two study periods comprised a 2-week run-in phase, followed by a 4-week blinded study-product consumption phase with intake of 4 g chicory inulin or maltodextrin (placebo) 3 times daily. Fortyfour healthy adult men and women with mild constipation (defined as 2–3 stools/week) documented their stool frequency and consistency, gastrointestinal characteristics and quality of life during the study.

Consumption of Orafti Inulin significantly increased stool frequency compared with placebo [median 4.0 (IQR 2.5–4.5) vs. 3.0 (IQR 2.5–4.0) stools/week, $P = 0.038$]. This was accompanied by a softening of stool consistency and trend towards higher satisfaction in comparison with placebo ($P = 0.059$). Tolerance of study products was excellent throughout the study. The beneficial effect of chicory inulin on bowel function, namely increased stool frequency, was achieved at a dosage of 12 g/day without resulting in gastrointestinal discomfort. Overall, a significant increase of bowel movements was documented with Orafti Inulin, which was accompanied by a softening of stool consistency and that had a positive impact on the quality of life, primarily by increasing satisfaction (Micka et al., 2016).

From faecal samples collected during this randomised controlled trial, faecal microbiota composition and metabolite profiles were determined using next-generation sequencing technology (16S rDNA microbiota profiling) and matching (gas chromatography mass spectrometry) metabolomics (Vandeputte et al., 2017). Orafti Inulin was shown to have a selective effect on the human gut microbiota, with specific changes to the relative abundances observed for *Bifidobacterium*, *Anaerostipes* and *Bilophila*. The observed decrease in abundances of *Bilophila*, a genus containing known pathobionts, following inulin consumption was associated with both softer stools and a favourable change in constipation-specific quality-of-life measures (Vandeputte et al., 2017).

This proprietary human intervention study, together with the accompanying microbiota and metabolite profiling analysis, extended the body of evidence on distinct health benefits related to prebiotic chicory root-derived inulin and provided further insights into the mechanism of action.

10.4 Summary of substantiating evidence

Inulin from chicory is a fermentable dietary fibre. The claimed effect referred to the ability of the fully fermentable dietary fibre chicory inulin to increase the frequency of bowel movements and thus to improve bowel function. Changes in bowel function such as more frequent bowel movements, increased faecal bulk, softer stools or reduced transit time have been considered beneficial physiological effects, provided they do not result in diarrhoea (EFSA, 2011).

The evidence submitted to EFSA for the substantiation of the claimed effect, following a detailed literature review, included 10 pertinent human intervention studies investigating the relationship between chicory inulin and improvement in bowel function by assessing the frequency of bowel movements (Table 10.1). In addition to the proprietary study by Schulz and Schön (2012) (published in Micka et al., 2016), EFSA regarded seven further studies as suitable for drawing conclusions for the substantiation of the claim (Bouhnik et al., 2007; Den Hond et al., 2000; Gibson et al., 1995; Gråsten et al., 2003; Kleessen et al., 1997, 2007; Kolida et al., 2007). Two studies were considered by the NDA Panel as not sufficient for drawing scientific conclusions, as they were only available as a poster abstract (Isakov et al., 2013) or examined a product with a low degree of polymerisation (Brighenti et al., 1999).

The pertinent studies compared the actual number of bowel movements (stool frequency) following the intake of inulin from chicory with the bowel movement frequency observed after intake of a reference (e.g. wheat pentosan as the major constituent of

Table 10.1 Strengths of substantiating evidence

Article 13.5 application on Orafti Inulin and bowel frequency: strengths of the evidence
Mechanism of action described in detail
Effect shown starting from 3 g inulin/day
More pronounced effects with significant levels at 12 g inulin/day
Effect shown for different study durations (1–4 weeks)
Effect shown in different study groups (normal/reduced stool frequency, irritable bowel syndrome with constipation)
Effect shown under different conditions (free living, clinical setting and metabolic suite)
Effect shown with different food applications (liquid, solid and semi-solid)
No data contradicting the claimed effect

wheat bran fibre), a placebo control (e.g. maltodextrin and sucrose) or an intervention without inulin. Studies comprised consumption periods from 1- to 4-week duration. The pertinent human intervention studies were conducted in male and female adult subjects of a broad age range from 19 to 89 years and with a body weight in the normal weight-to-overweight range (BMI < 30 kg/m^2). They were carried out at different locations in Europe including Belgium, England, Finland, France, Italy and Germany. Study groups comprised persons with normal bowel movement frequency as well as constipated individuals. Thus, study groups could be considered as representative of the general population.

Further, the increase in bowel movement frequency was seen upon chicory inulin consumption via different food applications, including supplementation in drinks and solid food products, for example biscuits, bread, a snack bar or rice cereal. This confirms that chicory inulin increases bowel movement frequency irrespective of the nature of the food application through which it is administered and solely on the basis of the 'active principle', that is the actual inulin content contained in different food applications.

The claimed effect of chicory inulin on improving bowel function is obtained with continuing consumption of chicory inulin. A significant increase in bowel movement frequency upon chicory inulin ingestion has been already achieved after a 1-week intervention period (Den Hond et al., 2000) and this favourable and significant increase in bowel movement frequency has also been seen when chicory inulin is administered during prolonged periods (Micka et al., 2016). These data confirm that the effect of chicory inulin on increasing bowel movement frequency is sustainable over time and also maintained over longer periods with continued inulin consumption.

Hereafter, the main outcomes of individual human intervention studies are briefly summarised. The key human intervention study by Micka et al. (2016) has already been described in detail.

Den Hond et al. (2000) showed a significantly higher bowel movement frequency during chicory inulin supplementation in comparison with placebo (+63%, $P = 0.02$). Stool bulk measured as faecal wet and dry weights was increased by 24 and 18% respectively (n.s.). This corresponds to an increase in stool bulk of 1.5 g for each gram inulin consumed ('bulking index'). The study by Gråsten et al. (2003) found similar effects of chicory inulin on bowel movement frequency compared with wheat pentosan or even more pronounced improvements as indicated by a greater increase in bowel movement frequency (+40%, n.s.). Wheat pentosan is the major constituent of wheat bran fibre which has already been positively evaluated with respect to its effect on bowel function (EFSA, 2010a). Stool consistency measured as the proportion of solid and soft faeces was improved by inulin as well as by wheat pentosan (n.s.). Kleessen et al. (1997) demonstrated a higher bowel movement frequency upon ingestion of chicory inulin in comparison with lactose, a commonly used mild laxative agent (+51% for 20 g and +95% for 40 g). The percentage dry matter of faeces was reduced by all treatments, indicating a softening effect on stool consistency. Softer stools were reported in the inulin group. The study by Kleessen et al. (2007) showed a significant increase in bowel movement frequency upon inulin supplementation (+43%, $P < 0.05$). This resulted in higher bowel movement frequency at the end of the intervention in comparison with the placebo group (+11%). Stool consistency, measured on a scale from 1 (hard) to 4 (watery), indicated a significant softening of faeces upon inulin ingestion (+17%, $P < 0.05$), whereas in the placebo group no significant change was observed. Gibson et al. (1995) reported a higher bowel movement frequency in healthy volunteers during the chicory inulin period in comparison with placebo (+13%, n.s.). This difference did not reach statistical significance, possibly due to the very low number of subjects ($n = 4$). Stool bulk, as measured by faecal wet and dry weights, was higher upon inulin ingestion by 34 and 23% respectively (n.s.). This corresponds to an increase in stool weight of 2 g per gram inulin consumed ('bulking index'). This is comparable with other non-starch polysaccharides such as pectin (Cummings, 2001). Dry matter of faeces was lower upon inulin ingestion (−4.5%, n.s.) which indicates a softening of stools.

The corresponding data for bowel function parameters from these human intervention studies collectively show that chicory inulin improves bowel function by leading to more frequent bowel movements and thus contributes to the maintenance of normal bowel function. No data have been identified that contradict the claimed effect. The effect of chicory inulin increasing the bowel movement frequency is obtained without leading to diarrhoea. Accordingly, the NDA panel considered in their scientific opinion that 'taken collectively, these studies provide evidence for an increase in stool frequency with the consumption of "native chicory inulin" at doses of 12 g/day' and that 'no adverse effects were reported in any of the studies mentioned' (EFSA, 2015).

10.4.1 Mechanism of action

The mechanism by which chicory inulin improves bowel function was also described and explained in detail in the Article 13.5 health claim application. The effect is repeatable and biologically plausible. Consequently, EFSA acknowledged in its scientific opinion 'the plausible mechanisms by which inulin and inulin-type fructans in "native chicory inulin" could exert the claimed effect' (EFSA, 2015).

Fig. 10.1 provides a graphic illustration of the mechanism of action of chicory inulin on bowel function. Chicory inulin is a non-digestible carbohydrate and fully fermentable. Due to the $\beta(2\rightarrow1)$ glycosidic bonds, inulin-type fructans resist hydrolysis and absorption in the human small intestine and reach the large intestine essentially complete. There they are selectively fermented to short-chain fatty acids, lactate and gases by colonic bacteria (Roberfroid and Slavin, 2000). This is accompanied by an increase in bacterial cell mass and a higher water content of the digesta. Consequently, inulin-type fructans lead to softer stools and facilitate excretion and also lead to enhanced propulsion of colonic contents via chemical and mechanical stimulation of the peristaltic reflex, resulting in an increase in the frequency of bowel movements (Micka et al., 2016).

Six human studies were provided in the health claim application as pertinent human mechanistic studies. These show the non-digestibility of inulin in the small intestine, its microbial fermentation in the large intestine and its influence on stool consistency, stool bulk or transit time. In other words, these studies investigated the mechanism by which inulin improves bowel function (Brighenti et al., 1995; Castiglia-Delavaud

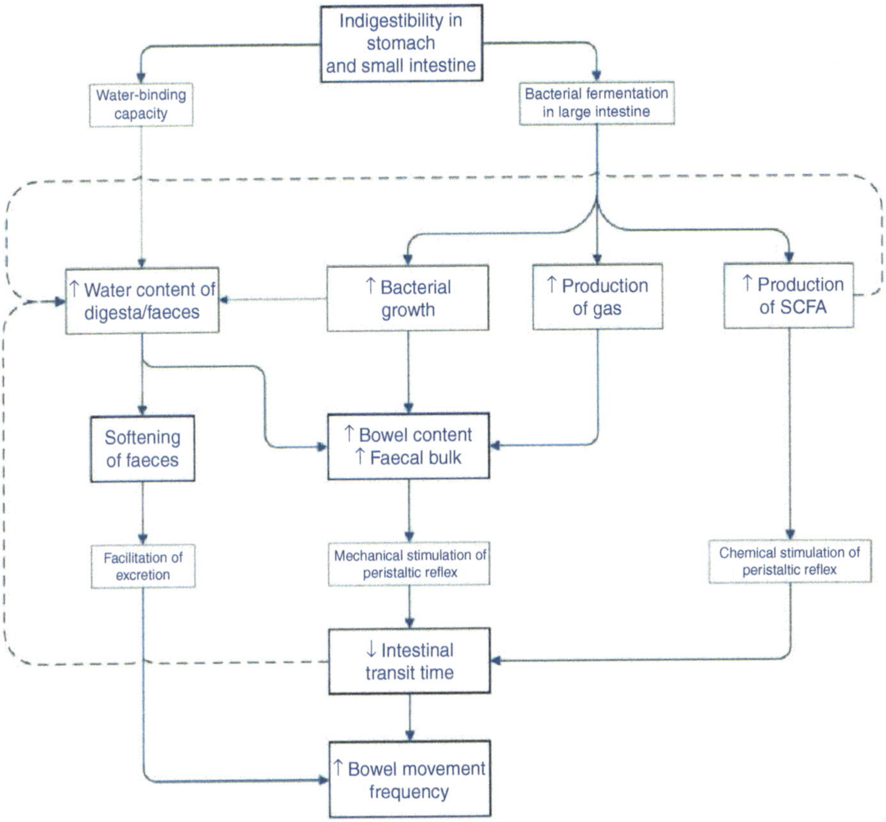

Figure 10.1 Mechanism of action of chicory inulin on bowel function and laxation. *SCFA*, Short-chain fatty acid.

et al., 1998; Ellegård et al., 1997; Marteau et al., 2011; Rumessen and Gudmand-Høyer, 1998; Van Dokkum et al., 1999). In addition, various studies in different animal species (such as dogs, cats, goats, rats and pigs) as well as in vitro studies were included in the application. These reported higher stool bulk and softer stools upon ingestion of inulin-type fructans, or provided evidence related to the production of short-chain fatty acids by bacterial degradation of inulin (Delzenne et al., 1995; Diez et al., 1998; Hesta et al., 2001; Kara et al., 2012; Kleessen et al., 2001; Pierce et al., 2005; Rycroft et al., 2001; Timm et al., 2010; Verlinden et al., 2006; Wang and Gibson, 1993). These data from pertinent human intervention and mechanistic studies consistently prove the cause-and-effect relationship between the consumption of inulin and improvement in bowel function as determined by more frequent bowel movements.

In addition to the EFSA evaluation, a recently published meta-analysis of five intervention studies has demonstrated significant effects of inulin on stool frequency, stool consistency and transit time, concluding that inulin intake has a positive effect on bowel function (Collado Yurrita et al., 2014). Further research on digestive health benefits with prebiotic inulin-type fructans also continues for other population groups, including young children (Closa-Monasterolo et al., 2016).

10.5 Importance of the claim in product development and consumer issues

In line with the recognised health benefits of dietary fibre consumption, EFSA considered the beneficial physiological effects on bowel function as the basis for the adequate intake of dietary fibre and set a level of 25 g/day for adults (adequate for normal laxation) (EFSA, 2010b). When comparing actual fibre intake to the recommended levels in the respective countries, it appears that dietary fibre intake does not meet the recommendations in many countries and regions (Moser et al., 2015). Thus there is an impressive 'fibre gap' despite the widespread knowledge of the health benefits of dietary fibres (Moser et al., 2015). To achieve the dietary goal of adequate fibre intake on an individual level as well as by the general population, consumption of more fibre-containing foods and foods enriched with fibres such as inulin seems to be desirable.

Chicory root-derived inulin possesses not only unique nutritional characteristics, but also technological characteristics and advantages, allowing its easy implementation as an ingredient into foods. Accordingly, it can be used in numerous product applications to replace fat or sugar, to improve texture and/or to increase the fibre content of food applications without modifying delicate flavours (Moser and Wouters, 2014). The proprietary health claim for inulin and bowel function can be used in all categories of the food supply. Example applications could be yoghurt, yoghurt drink, muesli bar, meal or meal replacement products.

'Digestive wellness' is considered as the 'key trend number 1' among the '10 key trends in food, nutrition and health for 2017' owing to new technologies and new understanding of the effects that gut health has on overall health (Mellentin, 2016). Quantitative consumer research on digestive health conducted on behalf of BENEO aimed to gain further consumer insights into how important and relevant digestive health is to consumers. This research shows that consumers see digestive health as key to their health and well-being (BENEO, 2016). Nearly three out of four European consumers eat products supporting digestive health because they feel it is good for their overall health. Moreover, 'digestive health claims' are seen as more convincing than 'high-in-fibre' claims only. The vast majority of consumers find that the authorised health claim adds clarity and credibility to the digestive health message (BENEO, 2016).

Incorporating inulin into food and beverage applications in combination with the health claim can help food manufacturers to meet consumer demand for products that support digestive health. Moreover, by using the health claim, food producers can help consumers to understand and benefit from the positive effect chicory inulin has on bowel function. The authorised health claim could be combined with general health-related well-being claims under Article 10.3. According to the European Commission implementing decision of 24 January 2013, Article 10.3 allows the use of easy, attractive statements which make reference to general, non-specific benefits of a food for overall good health or health-related wellbeing provided they are accompanied by a specific, authorised health claim. The use of such statements is regarded as helpful to consumers, as they would convey more consumer-friendly messages (2013/63/EU). Such messages, amongst others, could include: 'chicory inulin promotes digestive health' and 'chicory root fibre supports a healthy and balanced digestive system'.

Manufacturers now have considerable opportunities to build nutritional functionality into their products and communicate these benefits to the consumer, based on the authorised proprietary health claim for chicory inulin improving bowel function.

10.6 Conclusions

In general, inulin and inulin-type fructans are dietary fibres with established prebiotic effects. The specific changes both in composition and/or activity of the gastrointestinal microbiota provide benefits for human health. Physiological health benefits of chicory-derived inulin-type fructans include the positive modulation of gut microbiota, effects on energy intake and body weight, mineral absorption and bone health as well as on laxation and bowel function. These benefits have been intensively researched over more than 20 years. High-quality nutrition research has also created the foundation for the authorised EU health claim for chicory inulin and its effect of improving digestive health. The authorised Article 13.5 health claim on bowel function (increased stool frequency) benefits with chicory inulin is one of only few examples of health claims for dietary fibres approved till now under the EU Nutrition and Health Claims Regulation and is the first health claim authorised for a prebiotic fibre. With this proprietary health claim, food manufacturers are offered the opportunity to help consumers make better choices in their daily diet to increase their dietary fibre intake and support digestive health.

References

BENEO, 2016. New research shows digestive health is key to wellbeing for European consumers. (BENEO conducted online quantitative consumer research in the United Kingdom, Spain, France and Germany. The survey was carried out in 2015 and 2016 by Insites, a consumer research agency with offices in Belgium, the Netherlands, United Kingdom and the United States.) Available from: http://www.beneonews.com/Press_Releases/2016/Consumer_Research_Digestive_Health/BENEO_ConsumerResearchDigestiveHealth_2016.pdf.

Bouhnik, Y., Raskine, L., Champion, K., Andrieux, C., Penven, S., Jacobs, H., Simoneau, S., 2007. Prolonged administration of low-dose inulin stimulates the growth of bifidobacteria in humans. Nutr. Res. 27 (4), 187–193.

Brighenti, F., Casiraghi, M.C., Canzi, E., Ferrari, A., 1999. Effect of consumption of a ready-to-eat breakfast cereal containing inulin on the intestinal milieu and blood lipids in healthy male volunteers. Eur. J. Clin. Nutr. 53 (9), 726–733.

Brighenti, F., Casiraghi, M.C., Pellegrini, N., Riso, P., Simonetti, P., Testolin, G., 1995. Comparison of lactulose and inulin as reference standard for the study of resistant starch fermentation using hydrogen breath test. Ital. J. Gastroenterol. 27 (3), 122–128.

Candelli, M., Nista, E.C., Zocco, M.A., Gasbarrini, A., 2001. Idiopathic chronic constipation: pathophysiology, diagnosis and treatment. Hepatogastroenterology 48, 1050–1057.

Castiglia-Delavaud, C., Verdier, E., Besle, J.M., Vernet, J., Boirie, Y., Beaufrere, B., De Baynast, R., Vermorel, M., 1998. Net energy value of non-starch polysaccharide isolates (sugarbeet fibre and commercial inulin) and their impact on nutrient digestive utilization in healthy human subjects. Br. J. Nutr. 80 (4), 343–352.

Closa-Monasterolo, R., Ferré, N., Castillejo-DeVillasante, G., Luque, V., Gispert-Llaurado, M., Zaragoza-Jordana, M., Theis, S., Escribano, J., 2017. The use of inulin-type fructans improves stool consistency in constipated children. A randomised clinical trial: pilot study. Int. J. Food Sci. Nutr. 68 (5), 587–594.

Closa-Monasterolo, R., Gispert-Llaurado, M., Luque, V., Ferre, N., Rubio-Torrents, C., Zaragoza-Jordana, M., Escribano, J., 2013. Safety and efficacy of inulin and oligofructose supplementation in infant formula: results from a randomized clinical trial. Clin. Nutr. 32 (6), 918–927.

Collado Yurrita, Y.L., San Mauro Martín, I., Ciudad-Cabañas, M.J., Calle-Purón, M.E., Cabria, Hernández, 2014. Effectiveness of inulin intake on indicators of chronic constipation: a meta-analysis of controlled randomized clinical trials. Nutr. Hosp. 30 (2), 244–252.

Commission Regulation (EU) 2015/2314 of 7 December 2015 authorising a health claim made on foods, other than those referring to the reduction of disease risk and to children's development and health and amending Regulation (EU) No 432/2012. OJEU L 328, 12 December 2015 p. 46.

Cummings, J.H., 2001. The effect of dietary fiber on fecal weight and composition. In: Spiller, G.A. (Ed.), CRC Handbook of Dietary Fiber in Human Nutrition. third ed. CRC Press, Boca Raton, FL, pp. 183–252.

Cummings, J.H., Antoine, J.M., Azpiroz, F., Bourdet-Sicard, R., Brandtzaeg, P., Calder, P.C., Gibson, G.R., Guarner, F., Isolauri, E., Pannemans, D., Shortt, C., Tuijtelaars, S., Watzl, B., 2004. PASSCLAIM—gut health and immunity. Eur. J. Nutr. 43 (Suppl. 2), II118–II173.

Delzenne, N., Aertssens, J., Verplaetse, H., Roccaro, M., Roberfroid, M., 1995. Effect of fermentable fructo-oligosaccharides on mineral, nitrogen and energy digestive balance in the rat. Life Sci. 57 (17), 1579–1587.

Den Hond, E., Geypens, B., Ghoos, Y., 2000. Effect of high performance chicory inulin on constipation. Nutr. Res. 20 (5), 731–736.

Diez, M., Hornick, J.L., Baldwin, P., Van Eenaeme, C., Istasse, L., 1998. The influence of sugarbeet fibre, guar gum and inulin on nutrient digestibility, water consumption and plasma metabolites in healthy Beagle dogs. Res. Vet. Sci. 64 (2), 91–96.

EFSA Panel on Dietetic Products, Nutrition, and Allergies (NDA), 2010a. Scientific Opinion on the substantiation of health claims related to wheat bran fibre and increase in faecal bulk (ID 3066), reduction in intestinal transit time (ID 828, 839, 3067, 4699) and contribution to the maintenance or achievement of a normal body weight (ID 829) pursuant to Article 13(1) of Regulation (EC) No. 1924/2006. EFSA J. 8 (10), 1817.

EFSA Panel on Dietetic Products, Nutrition, and Allergies (NDA), 2010b. Scientific opinion on dietary reference values for carbohydrates and dietary fibre. EFSA J. 8 (3), 1462.

EFSA Panel on Dietetic Products, Nutrition and Allergies (NDA), 2011. Guidance on the scientific requirements for health claims related to gut and immune function. EFSA J. 9 (4), 1984.

EFSA Panel on Dietetic Products, Nutrition and Allergies (NDA), 2015. Scientific Opinion on the substantiation of a health claim related to "native chicory inulin" and maintenance of normal defecation by increasing stool frequency pursuant to Article 13(5) of Regulation (EC) No. 1924/2006. EFSA J. 13 (1), 3951.

Ellegård, L., Andersson, H., Bosaeus, I., 1997. Inulin and oligofructose do not influence the absorption of cholesterol, or the excretion of cholesterol, Ca, Mg, Zn, Fe, or bile acid but increases energy excretion in ileostomy subjects. Eur. J. Clin. Nutr. 51 (1), 1–5.

Gibson, G.R., Beatty, E.R., Wang, X., Cummings, J.H., 1995. Selective stimulation of bifidobacteria in the human colon by oligofructose and inulin. Gastroenterology 108 (4), 975–982.

Gråsten, S., Liukkonen, K., Chrevatidis, A., El-Nezami, H., Poutanen, K., Mykkänen, H., 2003. Effects of wheat pentosan and inulin on the metabolic activity of fecal microbiota and on bowel function in healthy humans. Nutr. Res. 23 (11), 1503–1514.

Hesta, M., Janssens, G.P., Debraekeleer, J., De Wilde, R., 2001. The effect of oligofructose and inulin on faecal characteristics and nutrient digestibility in healthy cats. J. Anim. Physiol. Anim. Nutr. 85 (5–6), 135–141.

Institute of Medicine (IoM), 2005. Dietary Reference Intakes for Energy, Carbohydrate, Fiber, Fat, Fatty Acids, Cholesterol, Protein and Amino Acids. National Academies Press, Washington, DC.

Isakov, V., Pilipenko, V., Shakhovskaya, A., Tutelyan, V., 2013. Efficacy of inulin enriched yogurt on bowel habits in patients with irritable bowel syndrome with constipation: a pilot study. FASEB J. 27, lb426.

Kara, C., Orman, A., Gencoglu, H., Kovanlikaya, A., Meral, Y., Cetin, I., Yibar, A., Kasap, S., Turkmen, I., Deniz, G., 2012. Effects of inulin supplementation on selected faecal characteristics and health of neonatal Saanen kids sucking milk from their dams. Animal 6 (12), 1947–1954.

Kleessen, B., Hartmann, L., Blaut, M., 2001. Oligofructose and long-chain inulin: influence on the gut microbial ecology of rats associated with a human faecal flora. Br. J. Nutr. 86 (2), 291–300.

Kleessen, B., Schwarz, S., Boehm, A., Fuhrmann, H., Richter, A., Henle, T., Krueger, M., 2007. Jerusalem artichoke and chicory inulin in bakery products affect faecal microbiota of healthy volunteers. Br. J. Nutr. 98 (3), 540–549.

Kleessen, B., Sykura, B., Zunft, H., Blaut, M., 1997. Effects of inulin and lactose on fecal microflora, microbial activity, and bowel habit in elderly constipated persons. Am. J. Clin. Nutr. 65 (5), 1397–1402.

Kolida, S., Meyer, G., Gibson, G.R., 2007. A double-blind placebo-controlled study to establish the bifidogenic dose of inulin in healthy humans. Eur. J. Clin. Nutr. 61 (10), 1189–1195.

Marteau, P., Jacobs, H., Cazaubiel, M., Signoret, C., Prevel, J.M., Housez, B., 2011. Effects of chicory inulin in constipated elderly people: a double blind controlled trial. Int. J. of Food Sci. Nutr. 62 (2), 164–170.

Micka, A., Siepelmeyer, A., Holz, A., Theis, S., Schön, C., 2017. Effect of consumption of chicory inulin on bowel function in healthy subjects with constipation: a randomized, double-blind, placebo-controlled trial. Int. J. Food Sci. Nutr. 68 (1), 82–89.

Mellentin, J., 2016. 10 Key Trends in Food, Nutrition & Health 2017. New Nutrition Business.

Moser, M., Sentko, A., Alexiou, A., 2015. Inulin and health benefits. In: Ramawat, K.G., Merillon, J.M. (Eds.), Polysaccharides: Bioactivity and Biotechnology. Springer International Publishing, Switzerland, pp. 675–715.

Moser, M., Wouters, R., 2014. Nutritional and technological benefits of inulin-type oligosaccharides. In: Moreno, F.J., Sanz, M.L. (Eds.), Food Oligosaccharides: Production, Analysis and Bioactivity. John Wiley & Sons, Ltd., Chichester, United Kingdom.

Pierce, K.M., Sweeney, T., Brophy, P.O., Callan, J.J., McCarthy, P., O'Doherty, J.V., 2005. Dietary manipulation post weaning to improve piglet performance and gastro-intestinal health. Anim. Sci. 81 (3), 347–356.

Regulation (EC) No 1924/2006 of the european parliament and of the council of 20 December 2006 on nutrition and health claims made on foods ELI: http://data.europa.eu/eli/reg/2006/1924/2012-11-29.

Roberfroid, M.B., Slavin, J., 2000. Nondigestible oligosaccharides. Crit. Rev. Food Sci. 40 (6), 461–480.

Rumessen, J.J., Gudmand-Høyer, E., 1998. Fructans of chicory: intestinal transport and fermentation of different chain lengths and relation to fructose and sorbitol malabsorption. Am. J. Clin. Nutr. 68 (2), 357–364.

Rycroft, C.E., Jones, M.R., Gibson, G.R., Rastall, R.A., 2001. A comparative in vitro evaluation of the fermentation properties of prebiotic oligosaccharides. J. Appl. Microbiol. 91, 878–887.

Schulz, A., Schön, C., 2012. Effect of consumption of inulin on bowel motor function in subjects with constipation (unpublished study report).

Timm, D.A., Stewart, M.L., Hospattankar, A., Slavin, J.L., 2010. Wheat dextrin, psyllium, and inulin produce distinct fermentation patterns, gas volumes, and short-chain fatty acid profiles in vitro. J. Med. Food 13 (4), 961–966.

Vandeputte, D., Falony, G., Vieira-Silva, S., Wang, J., Sailer, M., Theis, S., Verbeke, K., Raes, J., 2017. Prebiotic inulin-type fructans induce specific changes in the human gut microbiota. Gut., doi: 10.1136/gutjnl-2016-313271. (Epub ahead of print).

Van Dokkum, W., Wezendonk, B., Srikumar, T.S., van den Heuvel, E.G.H.M., 1999. Effects of nondigestible oligosaccharides on large-bowel functions, blood lipid concentrations and glucose absorption in young healthy male subjects. Eur. J. Clin. Nutr. 53 (1), 1–7.

Verlinden, A., Hesta, M., Hermans, J.M., Janssens, G.P.J., 2006. The effects of inulin supplementation of diets with or without hydrolysed protein sources on digestibility, faecal characteristics, haematology and immunoglobulins in dogs. Br. J. Nutr. 96 (5), 936–944.

Wang, X., Gibson, G.R., 1993. Effects of the in vitro fermentation of oligofructose and inulin by bacteria growing in the human large intestine. J. Appl. Bacteriol. 75 (4), 373–380.

Further Reading

Deutsche Gesellschaft für Ernährung, 2015. Referenzwerte für die Nährstoffzufuhr. 2. Auflage, 1. Ausgabe.

Part Four

General Function Claims

Authorised EU health claim for chitosan

11

*Michele J. Sadler**, *Jess Funnell***
*Rank Nutrition Ltd, Bethersden, Kent, United Kingdom; **Bournemouth University, Poole, United Kingdom

11.1 Introduction

The authorised EU health claim for the ingredient chitosan relates to the maintenance of normal blood low-density lipoprotein-cholesterol (LDL-C) concentrations. Elevated levels of plasma LDL-C are causal to the process of atherosclerosis (ESC, 2016). Elevated LDL-C concentrations are considered to be >160 mg/dL (>4.14 mmol/L) (EFSA, 2011), at which risk of cardiovascular disease (CVD) is increased due to adverse effects on artery structure and functions.

LDL-C transports cholesterol from the liver to the extra hepatic tissues such as muscle and adipose tissue, and is often referred to as 'bad cholesterol'. High levels in the serum greatly increases the risk of developing atherosclerosis, a disease of the cardiovascular system in which the blood vessels become occluded by cholesterol-rich plaques (Nelson and Cox, 2013). Compelling evidence from multiple randomised controlled trials (RCTs) shows that reduction of total cholesterol (TC) and LDL-C can reduce the risk of developing CVD (ESC, 2011, 2016).

Multiple factors affect blood cholesterol levels including saturated fat intake, obesity and genetics. There is a well-established positive relationship between blood cholesterol and body mass index (BMI), with a graded increase in TC with increasing BMI (Garrow et al., 2000). According to recent estimates, over 50% of both men and women in the WHO European Region are overweight, and approximately 23% of women and 20% of men are obese (World Health Organization, 2014). Data on the prevalence of obesity and overweight in England is acquired from the Health Survey for England, an annual survey that explores lifestyle habits and the prevalence of certain health conditions to inform and to evaluate specific health-related policies (HSCIC, 2016). In 2014, data showed that 61.7% of adults in England were overweight or obese, the gender split being 65.3% of men and 58.1% of women (Public Health England, 2016). A raised BMI is associated with increased risk of CVD, diabetes and some cancers.

There is also a well-established positive relationship between blood cholesterol concentration and intake of saturated fat in the diet, such that lower intakes of saturated fats are associated with lower levels of LDL-C (Garrow et al., 2000). With most people in the UK over-consuming saturated fats, this remains a public health concern. Public health advice is that the average man should not consume more than 30 g of saturated fats a day and the average woman should not have more than 20 g of saturated fats

Foods, Nutrients and Food Ingredients with Authorised EU Health Claims. http://dx.doi.org/10.1016/B978-0-08-100922-2.00011-5
Copyright © 2018 Elsevier Ltd. All rights reserved.

a day (Public Health England, 2016). According to data from the Global Health Observatory, the global prevalence of raised TC among adults (≥5 mmol/L) is estimated to be 39% (37% for males and 40% for females) (World Health Organization, 2008).

When considering these statistics, identification of dietary ingredients or supplements such as chitosan that have the potential to aid reduction of cholesterol levels (mainly LDL-C) are likely to be beneficial for public health, provided that the consumer can be informed about its potential benefits. This chapter will discuss the relevance of the approved EU health claim for chitosan and the substantiating scientific evidence, and will consider both the commercial importance and consumer understanding of the claim.

11.2 The role of cholesterol and low-density lipoproteins (LDL) in CVD risk

Cholesterol can have both a vital and detrimental role in our health, depending on its form and concentration in the blood. It is essential for health, for example being a part of cell membranes and the precursor to steroid hormones and vitamin D synthesis. However, it does not need to be obtained from the diet due to the body's ability to synthesise it from simpler molecules (Erasmus, 1993).

LDL-C is predominantly composed of cholesterol and cholesteryl esters, unlike high-density lipoprotein cholesterol (HDL-C), which consists of protein-rich particles containing relatively little cholesterol or other lipids. HDL-C is thus denser and is the smaller of the lipoprotein particles. Its role is to carry cholesterol to the liver, and in contrast to LDL-C, a low concentration of HDL-C in the blood increases the risk of CVD.

When the sum of synthesised and dietary cholesterol exceeds that of body requirements, pathological accumulation of cholesterol can result in the development of atherosclerosis. Atherosclerosis is the amassing of LDL-C within an artery, forming plaques that can eventually occlude the artery completely.

Plaque formation begins when stressors cause trauma in the endothelium of arteries such as hypertension and toxins from cigarettes. LDL-C with partially oxidised fatty acyl groups (oxLDL-C) starts to adhere and accumulate within the injured epithelial cells lining the arteries. This also promotes the immune cells to cluster around these regions and as the cells engulf oxLDL-C molecules, these cells differentiate into macrophages, eventually turning into foam cells. (Nelson and Cox, 2013). OxLDL-C has been found to be the largest source of lipids within foam cells (Levitan et al., 2010).

Over time, the foam cells participate in inflammatory responses and tissue remodelling in the arterial intima (Valledor et al., 2015) including the development of scar tissue with extracellular material. Restriction of oxygenated blood to the brain or heart can trigger a myocardial infarction or stroke.

Reduction of circulating oxLDL-C in the blood is thought to be beneficial in reducing the risk of atherosclerotic effects in the body.

11.3 Relevance of the claim

CVD is a leading cause of death globally. In 2012, statistics showed that CVD caused 47% of all deaths in Europe and 40% in the European Union, with an estimated overall cost of €196 billion a year to the EU economy (ESC, 2012).

The nutritional approach for preventing CVD aims to reduce atherogenic and thrombogenic risk factors by reducing obesity, lowering TC and LDL-C, lowering blood pressure and reducing platelet aggregation and oxidative stress.

Data from the Framingham Study has been widely used to show how coronary heart disease (CHD) risk increases steadily with increasing levels of serum cholesterol (Garrow et al., 2000). Hence, there is a need for cholesterol reduction strategies amongst the general population, and a role for beneficial dietary ingredients and nutritional supplements to aid cholesterol reduction through dietary means.

11.4 Characterisation of chitosan

An important requirement for the approval of health claims is the adequate characterisation of the food constituent, in this case chitosan, a derivative of chitin.

After cellulose, chitin is the most abundant biopolymer on the earth. It is soluble in acidic aqueous media; however, it is insoluble in water. Properties of chitosan include high viscosity, polyelectrolyte behaviour, polyoxysalt formation, ability to form films and metal chelation. Although it is not derived from plants, chitosan is similar to dietary fibre that is indigestible by mammalian digestive enzymes. Due to chitosan's biocompatibility, biodegradability and low toxicity, it is used in multiple applications including food, cosmetics, biomedical and pharmaceutical functions (Rinaudo, 2006).

With reference to the European Food Safety Authority (EFSA), the scientific opinion of the Panel on Dietetic Products, Nutrition and Allergies (NDA) that assessed the health claim for chitosan, states that it is a linear, cationic polysaccharide containing randomly distributed β-(1-4)-linked D-glucosamine and *N*-acetyl-D-glucosamine (EFSA, 2011). It is created through the deacetylation of chitin, a tough, translucent amino polysaccharide that is the main constituent in the exoskeletons of arthropods such as the shells of crustaceans, the external coverings of insects and is also found in the cell walls of particular algae or fungi. The degree of deacetylation can be quantified by reputable methods, and ranges from 60% to 100% in commercial preparations, with the molecular weight of chitosan ranging from 3,800 to 20,000 Da in such preparations (EFSA, 2011). Presently, the most significant source of chitin and chitosan remains chemical processing of the waste fraction of the shellfish industry (Fig. 11.1).

11.5 Summary of substantiating evidence

EFSA considers the maintenance of normal blood LDL-C concentrations to be a beneficial physiological effect, as raised levels of LDL-C may compromise the normal structure and function of the arteries.

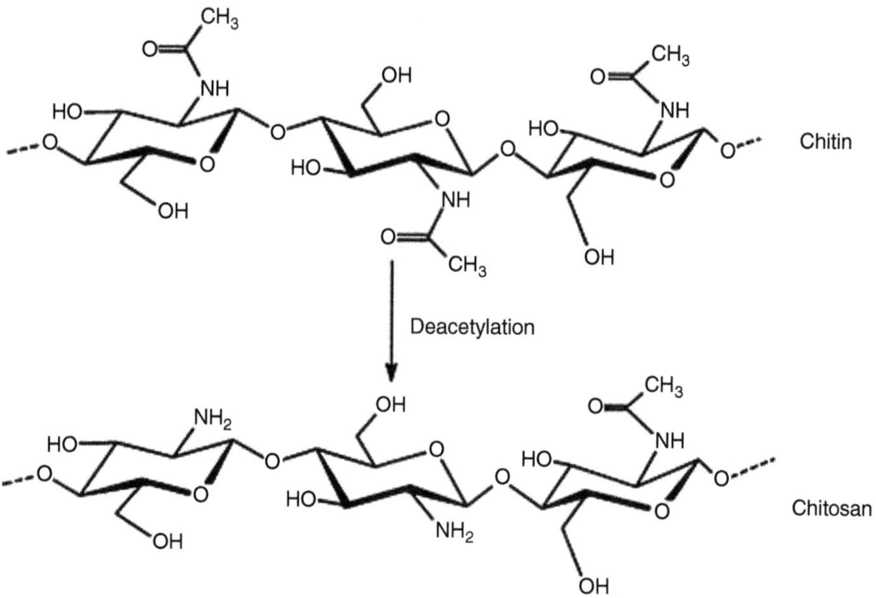

Figure 11.1 Chemical structures of chitin and chitosan.
Source: Zia, K.M., Barikani, M., Zuber, M., 2013. Advanced structured materials. Advances in Natural Polymers. Springer, Berlin, Heidelberg, pp. 55–119 (Figure 1).

Five animal studies and one human intervention study (Macchi, 1996) on the effects of chitosan on blood lipids were submitted for scientific substantiation of the claim. However, the NDA Panel noted that a Cochrane Systematic Review on chitosan (Jull et al., 2008), that had been submitted to support a claim for chitosan and reduction in body weight, had also analysed the effects of chitosan on blood cholesterol. Hence, the human studies identified in the Cochrane Systematic Review assessing the effects of chitosan on blood cholesterol were also considered by the NDA Panel in support of the claim.

The Cochrane Systematic Review included 15 RCTs lasting 4 or more weeks including a total of 1219 participants. With regard to body weight reduction, the analysis suggested that chitosan may have an effect on body weight, though the high-quality trials showed this effect to be minimal. Hence, a negative opinion was awarded for the body weight reduction claim.

However, nine trials in the meta-analysis provided data on TC levels (Colombo and Sciutto, 1996; Ho et al., 2001; Kaats et al., 2006; Macchi, 1996; Ni Mhurchu, 2003; Pittler et al., 1999; Veneroni et al., 1996; Wuolijoki et al., 1999; Zahorska-Markiewicz et al., 2002), and were therefore considered relevant to the cholesterol maintenance claim. As some of these studies used treatment preparations with other active ingredients in combination with chitosan, they were excluded by the NDA Panel on the basis that no scientific substation could be drawn due to uncertainty over which active ingredient was effecting cholesterol concentration.

The remaining studies that investigated the effects of chitosan intervention alone on blood cholesterol (Ho et al., 2001; Macchi, 1996; Ni Mhurchu, 2003; Pittler et al., 1999; Zahorska-Markiewicz et al., 2002), showed a small but statistically significant reduction in TC concentrations of −0.15 mmol/L (95 % CI −0.23 to −0.07, $P = 0.0002$). Analysis of only those trials that met the allocation concealment criteria, achieved similar results.

Seven of the trials in the meta-analysis included measurements of LDL-C concentrations (Colombo and Sciutto, 1996; Ho et al., 2001; Kaats et al., 2006; Ni Mhurchu, 2003; Veneroni et al., 1996; Wuolijoki et al., 1999; Zahorska-Markiewicz et al., 2002). However, combination treatments were used in four of these studies (Colombo and Sciutto, 1996; Kaats et al., 2006; Veneroni et al., 1996; Wuolijoki et al., 1999) and no other analysis of trials using chitosan in isolation was provided in the Cochrane Review.

The NDA Panel considered that the smaller studies with chitosan alone did not show a significant change in blood LDL-C concentrations; however, the largest study by Ni Mhurchu (2003), which included 250 subjects, detected a small but statistically significant decrease in LDL-C concentrations in support of chitosan (−0.12 mmol/L, 95% CI −0.19 to −0.05). When analysis was limited to the two studies with a duration of 6 months, similar results were obtained (−0.14 mmol/L, 95 % CI −0.19 to −0.06) (Ni Mhurchu et al., 2004; Zahorska-Markiewicz et al., 2002).

Seven trials provided data on HDL-C concentrations (Colombo and Sciutto, 1996; Ho et al., 2001; Kaats et al., 2006; Macchi, 1996; Ni Mhurchu, 2003; Veneroni et al., 1996; Zahorska-Markiewicz et al., 2002). Again, three of these trials tested other active ingredients in addition to chitosan (Colombo and Sciutto, 1996; Kaats et al., 2006; Veneroni et al., 1996), with no separate analysis of the trials using only chitosan provided in the review. Amongst the studies testing only chitosan, the study with the smallest sample size did show a statistically significant increase in HDL-C concentrations in comparison with placebo (Macchi, 1996). However, no significant differences were observed in the other three studies including the largest by Ni Mhurchu (2003).

Overall, the meta-analysis by Jull et al. (2008) showed that the consumption of 3 g of chitosan/day resulted in a small but statistically significant reduction of both TC (combining five studies) and LDL-C (combining two studies) concentrations, with no (undesirable) reduction observed in HDL-C concentrations.

11.6 Proposed mechanism for the effect

The mechanism suggested for the claimed effect is the process whereby the positive anions on chitosan bind to negatively charged lipids from dietary fats, which results in the reduction of their gastro-intestinal absorption and a theoretical decrease in total circulating blood lipids. This effect was observed in some animal studies with chitosan (Erasmus, 1993; Sugano et al., 1980; Zacour et al., 1992). However, in healthy humans the effects of chitosan on 24-h faecal fat excretion at doses of approximately 3 g/day were not statistically significant (Guerciolini et al., 2001), and therefore it remains unclear whether this could be a mechanism for the claimed effect.

In summary, the EFSA NDA Panel took into account that a meta-analysis of RCTs investigating the effects of chitosan consumption on blood lipids, showed overall a small but statistically significant reduction in TC and LDL-C concentrations (EFSA, 2011). The Panel concluded that a cause-and-effect relationship has been established between the consumption of chitosan and maintenance of normal blood LDL-C concentrations which is deemed by EFSA to be a beneficial physiological effect.

11.7 Authorised EU health claim for chitosan

11.7.1 Claim wording

For a health claim to be authorised, it must be agreed by the European Commission in conjunction with Member States and the Council, with approval of the European Parliament (Binns, 2014). Claim wording is an important aspect of discussion as the claim goes through the authorisation processes. It is formulated through careful consideration by the European Commission in collaboration with Member States and EFSA during the claim approval stages, to accurately reflect the scientific evidence for the food or food constituents. The official authorised wording of a claim is generally flexible when used in commercial communications such as the advertising and marketing of products, provided any changes made to the wording do not alter the meaning of the approved health claim or exaggerate the accepted beneficial effect.

The proposed wording for chitosan in the claim application was that it 'stimulates the regulation of cholesterol levels due to O-carboxymethyl chitosan'. In EFSA's scientific opinion on chitosan (EFSA, 2011), the proposed wording by the Panel was that 'Chitosan may contribute to maintaining normal blood cholesterol levels'.

Any changes to the wording suggested by EFSA are usually made to accommodate consumer understanding. In accordance with Article 13.1 of Regulation (EC) No. 1924/2006, the final authorised list of claims, published as Commission Regulation (EU) No. 432/2012 in the *Official Journal of the European Union* 25.5.12, is that 'Chitosan contributes to the maintenance of normal blood cholesterol levels'.

In the interests of consumer understanding, the authorised wording of the claim refers to cholesterol rather than LDL-C, in common with other authorised health claims relating to reduction or maintenance of blood cholesterol (Table 11.1).

11.7.2 Target population

The target population for the claim is adults.

11.7.3 Conditions of use

In accordance with the European Nutrition and Health Claims Regulation [(EC) No. 1924/2006], the subsequent information must also be incorporated in conjunction with use of the claim:

- A statement promoting the importance of a varied/balanced diet alongside a healthy lifestyle.

Table 11.1 Foods and ingredients with authorised EU health claims for the lowering/reduction and maintenance of cholesterol

Foods or ingredients	Claim wording	Conditions of use/restrictions of use of the claim	References
'Maintain cholesterol' (Article 13.1)			
ALA	ALA contributes to the maintenance of normal blood cholesterol levels	The claim may be used only for foods which are at least a source of ALA as referred to in the claim source of omega-3 fatty acids as listed in the Annex to Regulation (EC) No. 1924/2006. Information shall be given to the consumer that the beneficial effect is obtained with a daily intake of 2 g of ALA	Wahle (2015)
Beta-glucans	Beta-glucans contribute to the maintenance of normal blood cholesterol levels	The claim may be used only for foods which contains at least 1 g of beta-glucans from oats, oat bran, barley, barley bran or from mixtures of these sources per quantified portion. To bear the claim, information shall be given to the consumers that the beneficial effect is obtained with a daily intake of 3 g of beta-glucans from oats, oat bran, barley, barley bran or from mixtures of these beta-glucans.	Harland (2014)
Chitosan	Chitosan contributes to the maintenance of normal blood cholesterol levels	The claim may be used only for foods which provide a daily intake of 3 g of chitosan. To bear the claim, information shall be given to the consumers that the beneficial effect is obtained with a daily intake of 3 g of chitosan.	Current chapter
Foods with a low or reduced content of saturated fatty acids	Reducing consumption of saturated fats contributes to the maintenance of normal blood cholesterol levels	The claim may be used only for foods which are at least low in saturated fatty acids, as referred to in the claim low saturated fats or reduced saturated fatty acids as referred to in the claim reduced (name of nutrient) as listed in the Annex to Regulation (EC) No. 1924/2006.	Livingstone (2015)

(*Continued*)

Table 11.1 **Foods and ingredients with authorised EU health claims for the lowering/reduction and maintenance of cholesterol (*cont.*)**

Foods or ingredients	Claim wording	Conditions of use/restrictions of use of the claim	References
Glucomannan (konjac mannan)	Glucomannan contributes to the maintenance of normal blood cholesterol levels	The claim may be used only for foods which provide a daily intake of 4 g of glucomannan. To bear the claim, information shall be given to consumers that the beneficial effect is obtained with a daily intake of 4 g of glucomannan. *Warning* of choking to be given for people with swallowing difficulties or when ingesting with inadequate fluid intake; advice on taking with plenty of water to ensure substance reaches stomach.	Kuczora (2015)
Guar gum	Guar gum contributes to the maintenance of normal blood cholesterol levels	The claim may be used only for foods which provide a daily intake of 10 g of guar gum. To bear the claim, information shall be given to the consumers that the beneficial effect is obtained with a daily intake of 10 g of guar gum. *Warning* of choking to be given for people with swallowing difficulties or when ingesting with inadequate fluid intake; advice on taking with plenty of water to ensure substance reaches stomach.	Kuczora (2015)
HPMC	HPMC contributes to the maintenance of normal blood cholesterol levels	The claim may be used only for foods which provide a daily intake of 5 g of HPMC. To bear the claim, information shall be given to the consumers that the beneficial effect is obtained with a daily intake of 5 g of HPMC. *Warning* of choking to be given for people with swallowing difficulties or when ingesting with inadequate fluid intake; advice on taking with plenty of water to ensure substance reaches stomach.	Kuczora (2015)

LA	LA contributes to the maintenance of normal blood cholesterol levels	The claim may be used only for foods which provide at least 1.5 g of LA per 100 g and per 100 kcal. Information shall be given to consumers that the beneficial effect is obtained with a daily intake of 10 g of LA.	Wahle (2015)
Monascus purpureous (red yeast rice)	Monacolin K from red yeast rice contributes to the maintenance of normal blood cholesterol levels	The claim may be used only for foods which provide a daily intake of 10 mg of monacolin K from red yeast rice. To bear the claim, information shall be given to consumers that the beneficial effect is obtained with a daily intake of 10 mg of monacolin K from fermented red yeast rice preparations.	Le Bloc'h et al. (2015)
MUFAs and PUFAs	Replacing saturated fats with unsaturated fats in the diet contributes to the maintenance of normal blood cholesterol levels (MUFA and PUFA are unsaturated fats)	The claim may be used only for foods which are high in unsaturated fatty acids, as referred to in the claim high unsaturated fats as listed in the Annex to Regulation (EC) No. 1924/2006.	Livingstone (2014)
Oleic acid	Replacing saturated fats in the diet with unsaturated fats contributes to the maintenance of normal blood cholesterol levels. Oleic acid is an unsaturated fat	The claim may be used only for foods which are high in unsaturated fatty acids, as referred to in the claim high unsaturated fats as listed in the Annex to Regulation (EC) No. 1924/2006.	Livingstone (2014)
Pectins	Pectins contribute to the maintenance of normal blood cholesterol levels	The claim may be used only for foods which provide a daily intake of 6 g of pectins. To bear the claim information shall be given to consumers that the beneficial effect is obtained with a daily intake of 6 g of pectins. *Warning* of choking to be given for people with swallowing difficulties or when ingesting with inadequate fluid intake; advice on taking with plenty of water to ensure substance reaches stomach.	Paulionis et al. (2015)

(*Continued*)

Table 11.1 Foods and ingredients with authorised EU health claims for the lowering/reduction and maintenance of cholesterol (*cont.*)

Foods or ingredients	Claim wording	Conditions of use/restrictions of use of the claim	References
Plant sterols/stanols	Plant sterols/stanols contribute to the maintenance of normal blood cholesterol levels	To bear the claim information shall be given to the consumers that the beneficial effect is obtained with a daily intake of at least 0.8 g of plant sterols/stanols.	Shortt (2015)
Lower/reduce cholesterol (Article 14.1a)			
MUFAs/PUFAs	Replacing saturated fats with unsaturated fats in the diet has been shown to lower/reduce blood cholesterol. High cholesterol is a risk factor in the development of CHD	The claim may be used only for foods which are high in unsaturated fatty acids, as referred to in the claim high unsaturated fats as listed in the Annex to Regulation (EC) No. 1924/2006. The claim may only be used for fats and oils.	Livingstone (2017)
Barley beta-glucans	Barley beta-glucans have been shown to lower/reduce blood cholesterol. High cholesterol is a risk factor in the development of CHD	Information shall be given to consumers that the beneficial effect is obtained with a daily intake of 3 g of barley beta-glucan. The claim can be used for foods which provide at least 1 g of barley beta-glucan per quantified portion.	Harland (2014)
Oat beta-glucan	Oat beta-glucan has been shown to lower/reduce blood cholesterol. High cholesterol is a risk factor in the development of CHD	Information shall be given to consumers that the beneficial effect is obtained with a daily intake of 3 g of oat beta-glucan. The claim can be used for foods which provide at least 1 g of oat beta glucan per quantified portion.	Harland (2014)

Plant sterols/stanol esters	Plant sterols/stanol esters have been shown to lower/reduce blood cholesterol. High cholesterol is a risk factor in the development of CHD	Information to consumers that the beneficial effect is obtained with a daily intake of 1.5–3 g of plant sterols/stanols. Reference to the magnitude of the effect may only be made for foods within the following categories: yellow fat spreads, dairy products, mayonnaise and salad dressings. When referring to the magnitude of the effect, the range '7% to 10%' for foods that provide a daily intake of 1.5–2.4 g of plant sterols/stanols or the range '10% to 12.5%' for foods that provide a daily intake of 2.5–3 g of plant sterols/stanols and the duration to obtain the effect in '2 to 3 weeks' must be communicated to consumers.	Shortt (2015)
Plant sterols: sterols extracted from plants, free or esterified with food-grade fatty acids	Plant sterols have been shown to lower/reduce blood cholesterol. High cholesterol is a risk factor in the development of CHD	Information to consumers that the beneficial effect is obtained with a daily intake of 1.5–3 g of plant sterols. Reference to the magnitude of the effect may only be made for foods within the following categories: yellow fat spreads, dairy products, mayonnaise and salad dressings. When referring to the magnitude of the effect, the range '7% to 10%' for foods that provide a daily intake of 1.5–2.4 g of plant sterols or the range '10% to 12.5%' for foods that provide a daily intake of 2.5–3 g of plant sterols and the duration to obtain the effect 'in 2 to 3 weeks' must be communicated to consumers.	Shortt (2015)
Plant stanol esters	Plant stanol esters have been shown to lower/reduce blood cholesterol. High cholesterol is a risk factor in the development of CHD	Information to consumers that the beneficial effect is obtained with a daily intake of 1.5–3 g of plant stanols. Reference to the magnitude of the effect may only be made for foods within the following categories: yellow fat spreads, dairy products, mayonnaise and salad dressings. When referring to the magnitude of the effect, the range '7% to 10%' for foods that provide a daily intake of 1.5–2.4 g of plant stanols or the range '10%-12.5%' for foods that provide a daily intake of 2.5–3 g of plant stanols and the duration to obtain the effect 'in 2 to 3 weeks' must be communicated to the consumer.	Shortt (2015)

ALA, Alpha-linolenic acid; CHD, coronary heart disease; HPMC, hydroxypropyl methylcellulose; LA, linoleic acid; MUFA, mono-unsaturated fatty acid; PUFA, polyunsaturated fatty acid.

- The quantity of the food and pattern of consumption necessary to acquire the claimed benefit.
- A statement focused on consumers who should avoid using the food and any suitable warning for products that are likely to present a health risk if consumed in excess.

These specific conditions are set out in Article 10 of the Regulation, and are linked with the conditions of use. For chitosan, the EU register of health and nutrition claims states 'In order to bear the claim information shall be given to the consumer that the beneficial effect is obtained with a daily intake of 3 g of chitosan'. Therefore, the claim may be used only for products providing a daily intake of 3 g of chitosan.

With reference to risk of excess consumption, the NDA Panel did not find a need for any such statements in relation to chitosan, and therefore this information is not deemed necessary for products containing chitosan. Chitosan is generally considered non-toxic and non-digestible because unlike some medications for hyperlipidaemia, it does not affect enzymes in cholesterol biosynthesis and is not absorbed.

11.8 Consumers understanding and perception of authorised claims

Provided that they are scientifically substantiated, nutrition and health claims can enable consumers to make well-informed food choices to help maintain or improve beneficial physiological effects within the body. For health claims to be permitted for use in the European Union, it is essential that consumers correctly understand the claim including the conditions of use. This is because health claims can have the potential to misdirect consumers towards food choices that might be against their own interests. Regulation (EC) No. 1924/2006 states 'It is important that claims on foods can be understood by the consumer and it is appropriate to protect all consumers from misleading claims'. This Regulation identifies specific requirements for using the claim including specification of the amount required to achieve the claimed effect and a requirement that this quantity can reasonably be expected to be consumed. At the same time, the claim wording must also reflect the relevant scientific evidence, for example for the supplement in question 'Chitosan contributes to the maintenance of normal blood cholesterol levels'.

Consumers have their own preset ideas about diet and nutrition, such that a claim may bring to their mind ideas about other nutrients, for example 'low cholesterol' could mean to an individual consumer that the product may be 'low in fat'. This might lead the consumer to draw a conclusion that potentially goes beyond the claim, resulting in misinterpretation of the claim (Leathwood et al., 2007).

When consumers encounter new information such as that in a claim, they engage in some form of cognitive action where they link the new information to information previously stored in their memory, and this can sometimes result in inferences drawn about the product carrying the health claim that might go beyond what is clearly stated in the health claim (Grunert et al., 2011). Inferences such as the 'magic bullet' effect and the 'halo' effect have been widely studied in relation to health claims where a consumer believes that the product carrying the claim is healthier than it actually is (Andrews et al., 2000; Hieke and Grunert, 2017).

EU Regulation 1924/2006 requires that the 'average consumer' is able to understand nutrition and health claims. Recital 16 defines the average consumer as someone 'who is reasonably well informed and reasonably observant and circumspect, taking into account social, cultural and linguistic factors'. Influences related to consumers understanding include age, gender, culture, socio-economic status, education and knowledge of nutritional information (Nocella and Kennedy, 2012). For example middle-aged and elderly consumers tend to be considerably more health-oriented than younger consumers. This is because members of their immediate social environment or older people themselves are much more likely to be diagnosed with a non-communicable disease than the younger consumers (Bech-Larsen and Grunert, 2003; Siegrest et al., 2008). As the risk of CVD increases with age, it is likely that older consumers will be more interested than younger consumers in the health claim for chitosan. Also, as chitosan is a specialist supplement product that is not widely known and as chitosan is not naturally occurring in most food, it will take time to raise consumer awareness of the claim.

11.9 Commercial importance and potential of chitosan in product development

The food and agricultural sector of the United Kingdom makes up to 14% of the national employment (3.9 million people) and annual total consumer expenditure on food, beverages and catering is just under £200 billion, such that an estimated 11% of all household income is spent on food (DEFRA, 2015). The food and beverages industry is incredibly competitive, with many failed products, showing the need for innovative ideas to break into the market. With health being a key driver in consumer choice, products carrying health claims may increase intention to purchase. With this in mind, the development of products that meet conditions for authorised health claims may be desirable to manufacturers and even critical to compete within the marketplace.

Consumers are much more likely to accept functional ingredients with a well-established and largely appealing health image than functional ingredients to which they are unaccustomed (Bech-Larsen and Grunert, 2003; Menrad, 2003; Poulsen, 1999; Urala and Lähteenmäki, 2007). This means it is far easier to acquire consumer approval for a food enhanced with ingredients that are well known for their health benefits such as calcium and vitamin C, than it is for dietary components that are unknown to the majority of the general public. As future success for functional foods demands skills in food marketing as well as scientific documentation, food and pharmaceutical companies may possibly join forces to take advantage of complementary competences (Bech-Larsen and Scholderer, 2007).

Chitosan is mainly sold in health food outlets in the form of a dry capsule, predominantly in 250–500 mg potencies. It is sometimes mixed with other 'fat binders' also. As a food ingredient chitosan can impart viscosity, sometimes forming gels, and can be used as a thickening agent, for example in chocolate milk (Bray and Bouchard, 2014). Other applications of chitosan within the food industry include clarification of beverages (chitosan salts), as a texture control agent and as an emulsifying agent. Manufacturers in Japan produce dietary cookies, potato chips and noodles that are enriched with chitosan due to its cholesterol-lowering effect as well as chitosan-enriched vinegar products (Shahidi et al., 1999).

Though the target population for the claim is the general adult population, those who are more health conscious are more likely to purchase products that contain or are supplemented with chitosan. The authorised health claim for chitosan may resonate more with consumers if it is used on products that fit the image of cholesterol maintenance through other attributes, or that are in a form that consumers who purchase cholesterol maintenance products are familiar with. For example there are many popular cholesterol-lowering dairy products enriched with plant sterols and/or stanols or stanolesters. Dairy products enriched with chitosan could therefore be a potential development that might prove popular in the health food market and offer consumers an alternative choice.

However, should chitosan be added to food products, it needs to be considered whether it may alter palatability of the food or have any other undesirable effects. For example the addition of dietary fibre into foods has been associated with negative influences on mineral bioavailability, predominantly in high-risk populations (Bosscher et al., 2001; Claye et al., 1998). A recent study demonstrated that when chitosan is added to a food similar to yoghurt, both glucose and calcium availabilities are decreased and this effect is more noticeable than that produced by the plant fibre (Rodríguez et al., 2008).

Several studies in the past have demonstrated that yoghurt with visible fibre is commonly rejected or received a less positive response by consumers not only due to its unfamiliar appearance, but also because of a 'fibre-related off-taste' (Hashim et al., 2009; Sanz et al., 2008; Sendra et al., 2008). Incorporating chitosan into food products would require rigorous sensory analysis to ensure palatability and appeal to the consumer. However, individuals with increased nutritional knowledge may be willing to consume such functional foods, as a significant relationship has been shown between nutritional knowledge and healthiness perception and willingness to try functional foods. Consumers with a low level of nutritional knowledge were not concerned by the consumption of functional foods, while consumers with the most dietary knowledge were interested in the enhancement of healthy products with fibre or antioxidants (Ares et al., 2008).

Overall, it is uncertain whether chitosan would be better received if it were incorporated into food products particularly branded products that may command brand loyalty, or sold in a capsule form. There are many considerations including how to add the recommended daily intake of chitosan into products without altering its palatability. Another potential issue is marketing and differentiation from other functional foods especially those that can also claim 'contributes to the maintenance of normal blood cholesterol levels' such as plant stanols and sterols, beta-glucans, alpha-linolenic acid, glucomannan, pectins and guar gum (Table 11.1).

11.10 Conclusions

The authorisation of a European health claim is carried out by the European Commission in conjunction with the Member States, European Council and the European Parliament with consideration of scientific assessment by the EFSA NDA Panel. The Panel is responsible for assessing the evidence provided for the scientific substantiation of the health claim and in the case of chitosan concluded that there is sufficient evidence to substantiate the claim for 'contributes to the maintenance of normal blood cholesterol levels' with a daily intake of 3 g of chitosan.

Sale of chitosan as a dietary supplement, or its incorporation into food or beverage products, along with use of the associated health claim, could potentially be advantageous for public health due to the prevalence of elevated blood cholesterol, especially LDL-C, and it's link to CVD. When considering the use of chitosan in products, food and beverage manufacturers will need to take into consideration consumer perception of health claims and challenges that may occur during product development such as taste and textural issues. Incorporation of chitosan into products that already have a 'healthy image' in the consumers' mind may be more positively received and factors such as brand loyalty, knowledge of the functional ingredient and sensory perception will also be important.

References

Andrews, J.C., Burton, S., Netemeyer, R.G., 2000. Are some comparative nutrition claims misleading? The role of nutrition knowledge, and claim type and disclosure conditions. J. Advert. 29 (3), 29–42.

Ares, G., Gimenez, A., Gambaro, A., 2008. Influence of nutritional knowledge on perceived healthiness and willingness to try functional foods. Appetite 51, 663–668.

Bech-Larsen, T., Grunert, K.G., 2003. The perceived healthiness of functional foods: a conjoint study of Danish, Finnish and American consumers' perception of functional foods. Appetite 40 (1), 9–14.

Bech-Larsen, T., Scholderer, J., 2007. Functional foods in Europe: consumer research, market experiences and regulatory aspects. Trends Food Sci. Technol. 18 (4), 231–234.

Binns, N., 2014. The regulation of health claims in Europe. Sadler, M.J. (Ed.), Foods, Nutrients and Food Ingredients With Authorised EU Health Claims, vol. 1, Woodhead Publishing, Cambridge, United Kingdom, pp. 3–22, (Chapter 1).

Bosscher, D., Van Ciaillie-Bertrand, M., Deelstra, H., 2001. Effect of thickening agents, based on soluble dietary fibre on the availability of calcium, iron, and zinc from infant formulas. Nutrition 17, 614–618.

Bray, G.A., Bouchard, C., 2014. Handbook of Obesity–Volume 2: Clinical Applications, fourth ed. CRC Press, USA.

Claye, S.S., Idouraine, A., Weber, C.W., 1998. In vitro mineral binding capacity of five fibre sources and their insoluble components for magnesium and calcium. Food Chem. 61, 333–338.

Colombo, P., Sciutto, A.M., 1996. Nutritional aspects of chitosan employment in hypocaloric diet. Acta Toxicol. Therapeut. 17 (4), 287–302.

Department for Environment, Food and Rural Affairs (DEFRA), 2015. Food Statistics Pocketbook 2015-In year update. Department for Environment, Food and Rural Affairs, United Kingdom. Available from: https://www.gov.uk/government/uploads/system/uploads/attachment_data/file/526395/foodpocketbook-2015update-26may16.pdf.

EFSA, 2011. Scientific opinion on the substantiation of health claims related to chitosan and reduction in body weight (ID 679, 1499), maintenance of normal blood LDL-cholesterol concentrations (ID 4663), reduction of intestinal transit time (ID 4664) and reduction of inflammation (ID 1985) pursuant to Article 13(1) of Regulation (EC) No. 1924/20061. Panel on Dietetic Products, Nutrition and Allergies (NDA) 2, 3 European Food Safety Authority (2011)Effect of the viscosity or deacetylation degree of chitosan on fecal fat excreted from rats fed on a high fat diet. EFSA J. 9, 2214.

Erasmus, U., 1993. Fats That Heal, Fats That Kill: The Complete Guide to Fats, Oils, Cholesterol and Human Health. Alive Books, Canada.

European Society of Cardiology (ESC), 2011. ESC/EAS Guidelines for the management of dyslipidaemias: The Task Force for the management of dyslipidaemias of the European Society of Cardiology (ESC) and the European Atherosclerosis Society (EAS). Eur. Heart J. 32, 1769–1818.

European Society of Cardiology (ESC), 2012. 2012 European Cardiovascular Disease Statistics. European Society of Cardiology, Europe. Available from: http://www.escardio.org/The-ESC/Initiatives/EuroHeart/2012-European-Cardiovascular-Disease-Statistics.

European Society of Cardiology (ESC), 2016. 2016 European Guidelines on cardiovascular disease prevention in clinical practice. Eur. Heart J. 37, 2315–2381.

Garrow, J.S., James, W.P.T., Ralph, A., 2000. Human Nutrition and Dietetics, tenth ed. Churchill, Livingstone, Edinburgh.

Grunert, K.G., Scholderer, J., Rogeaux, M., 2011. Determinants of consumer understanding of health claims. Appetite 56 (2), 269–277.

Guerciolini, R., Radu-Radulescu, L., Boldrin, M., Dallas, J., Moore, R., 2001. Comparative evaluation of fecal fat excretion induced by orlistat and chitosan. Obes. Res. 9 (6), 364–367.

Harland, J., 2014. Authorised EU health claims for barley and oat beta-glucans. Sadler, M.J. (Ed.), Foods, Nutrients and Food Ingredients with Authorised EU Health Claims, vol. 1, Woodhead Publishing, Cambridge, United Kingdom, (Chapter 2).

Hashim, I., Khalil, A., Afifi, H., 2009. Quality characteristics and consumer acceptance of yogurt fortified with date fiber. J. Dairy Sci. 92 (11), 5403–5407.

Hieke, S., Grunert, K.G., 2017. Introduction: consumers and health claims. Sadler, M.J. (Ed.), Foods, Nutrients and Food Ingredients with Authorised EU Health Claims, vol. 3, Woodhead Publishing, Cambridge, United Kingdom.

Ho, S.C., Tai, E.S., Eng, P.H.K., Tan, C.E., Fok, A.C.K., 2001. In the absence of dietary surveillance, chitosan does not reduce plasma lipids or obesity in hypercholesterolemic obese Asian subject. Singapore Med. J. 42 (1), 6–10.

Health and Social Care Information Centre (HSCIC), 2016. Health Survey for England; Health, Social Care and Lifestyles. Health and Social Care Information Centre, United Kingdom. Available from: http://www.hscic.gov.uk/article/3741/Health-Survey-for-England-Health-social-care-and-lifestyles.

Jull, A.B., Ni Mhurchu, C., Bennett, D.A., Dunshea-Mooij, C.A.E., Rodgers, A., 2008. Chitosan for overweight or obesity. Cochrane Database Syst. Rev. 16 (3), CD003892.

Kaats, G., Michalek, J.E., Preuss, H.G., 2006. Evaluating efficacy of a chitosan product using a double-blinded, placebo-controlled protocol. J. Am. Coll. Nutr. 25 (5), 389–394.

Kuczora, S., 2015. Authorised EU health claims for glucomannan, guar gum and hydroxymethylcellulose. Sadler, M.J. (Ed.), Foods, Nutrients and Food Ingredients with Authorised EU Health Claims, vol. 2, Woodhead Publishing, Cambridge, United Kingdom, (Chapter 10).

Le Bloc'h, J., Pauquai, T., Bourges, C., 2015. Authorised EU health claims for red yeast rice. Sadler, M.J. (Ed.), Foods, Nutrients and Food Ingredients with Authorised EU Health Claims, vol. 2, Woodhead Publishing, Cambridge, United Kingdom, (Chapter 8).

Leathwood, P.D., Richardson, D.P., Sträter, P., Todd, P.M., van Trijp, H.C.M., 2007. Consumer understanding of nutrition and health claims: sources of evidence. Br. J. Nutr. 98 (3), 474–484.

Levitan, I., Volkov, S., Subbaiah, P.V., 2010. Oxidized LDL: diversity, patterns of recognition, and pathophysiology. Antiox. Redox Signal. 13 (1), 39–75.

Livingstone, K.M., 2014. Authorised EU health claims for the replacement of saturated fats. Sadler, M.J. (Ed.), Foods, Nutrients and Food Ingredients with Authorised EU Health Claims, vol. 1, Woodhead Publishing, Cambridge, United Kingdom, (Chapter 14).

Livingstone, K.M., 2015. Authorised EU health claims for foods with a low or reduced content of saturated fats. Sadler, M.J. (Ed.), Foods, Nutrients and Food Ingredients with Authorised EU Health Claims, vol. 2, Woodhead Publishing, Cambridge, United Kingdom, (Chapter 14).

Livingstone, K.M., 2017. Authorised EU health claims for MUFA and PUFA. Sadler, M.J. (Ed.), Foods, Nutrients and Food Ingredients with Authorised EU Health Claims, vol. 3, Woodhead Publishing, Cambridge, United Kingdom, (Chapter 5).

Macchi, G., 1996. A new approach to the treatment of obesity: chitosan's effects on body weight reduction and plasma cholesterol levels. Acta Toxicol. Therapeut. 17 (4), 301–320.

Menrad, K., 2003. Market and marketing of functional foods. Eur. J. Food Eng. 56, 181–188.

Nelson, D.L., Cox, M.M., 2013. Lehninger Principles of Biochemistry, sixth ed. W.H. Freeman & Company, New York, NY.

Ni Mhurchu, C., 2003. The effect of the dietary supplement, chitosan, on body weight: a randomised controlled trial in 250 overweight and obese adults. Proc. Nutr. Soc. N. Eur. J. Food Eng. 28, 143.

Ni Mhurchu, C., Poppitt, S., McGill, A.-T., Leahy, F., Bennett, D., Lin, R., Ormrod, D., Ward, L., Strik, C., Rodgers, A., Collaboration, 2004. The effect of the dietary supplement, chitosan, on body weight: a randomised controlled trial in 250 overweight and obese adults. Int. J. Obes. Relat. Metab. Disord. 28, 1149–1156.

Nocella, G., Kennedy, O., 2012. Food health claims–what consumers understand? Food Pol. 37 (5), 571–580.

Paulionis, L., Walters, B., Li, K., 2015. Authorised EU health claims on pectins. Sadler, M.J. (Ed.), Foods, Nutrients and Food Ingredients with Authorised EU Health Claims, vol. 2, Woodhead Publishing, Cambridge, United Kingdom, (Chapter 9).

Pittler, M.H., Abbot, N.C., Harkness, E.F., Ernst, E., 1999. Randomized, double-blind trial of chitosan for body weight reduction. Eur. J. Clin. Nutr. 53, 379–381.

Poulsen, J., 1999. Danish consumers' attitudes towards functional foods. MAPP Working Paper No. 62, Aarhus School of Business, Aarhus.

Public Health England (PHE), 2016. The Eatwell Guide. Public Health England, United Kingdom. Available from: https://www.gov.uk/government/uploads/system/uploads/attachment_data/file/528200/Eatwell_guide_booklet.pdf.

Rinaudo, M., 2006. Chitin and chitosan: properties and applications. Prog. Poly. Sci. 31 (7), 603–632.

Rodríguez, M.S., Montero, M., Staffolo, M.D., Martino, M., Bevilacqua, A., Albertengo, L., 2008. Chitosan influence on glucose and calcium availability from yogurt: in vitro comparative study with plants fibre. Carbohydr. Polym. 74 (4), 797–801.

Sanz, T., Salvadore, A., Jimenez, A., Fiszman, S.M., 2008. Yogurt enrichment with functional asparagus fibre. Effect of fibre extraction method on rheological properties, colour, and sensory acceptance. Eur. Food Res. Technol. 227 (5), 1515–1521.

Sendra, E., Fayos, P., Lario, Y., Fernandez-Lopez, J., Savas-Barbara, E., Perez-Alvarez, J., 2008. Incorporation of citrus fibers in fermented milk containing probiotic bacteria. Food Microbiol. 25 (1), 13–21.

Shahidi, F., Arachchi, J.K.V., Jeon, Y., 1999. Food applications of chitins and chitosans. Trends Food Sci.Technol. 10 (2), 37–51.

Shortt, C., 2015. Authorised EU health claims for phytosterols. Sadler, M.J. (Ed.), Foods, Nutrients and Food Ingredients with Authorised EU Health Claims, vol. 2, Woodhead Publishing, Cambridge, United Kingdom, (Chapter 2).

Siegrest, M., Stampfli, N., Kastenholz, H., 2008. Consumers' willingness to buy functional foods. The influence of carrier, benefit and trust. Appetite 51 (3), 526–529.

Sugano, M., Fujikawa, T., Hiratsuji, Y., Nakashima, K., Fukuda, N., Hasegawa, Y., 1980. A novel use of chitosan as a hypocholesterolemic agent in rats. Am. J. Clin. Nutr. 33, 787–793.

Urala, N., Lähteenmäki, L., 2007. Consumers' changing attitudes towards functional foods. Food Qual. Pref. 18, 1–12.
Valledor, A.F, Lloberas, J., Celada, A, 2015. Macrophage Foam Cells. John Wiley & Sons Ltd., Chichester, DOI: 10.1002/9780470015902.a0020730.pub2.
Veneroni, G., Veneroni, F., Contos, S., Tripodi, S., De Bernardi, M., Guarino, C., Marletta, M., 1996. Effect of a new chitosan dietary integrator and hypocaloric diet on hyperlipidemia and overweight in obese patients. Acta Toxicol. Therapeut. 17 (1), 53–70.
Wahle, K.W.J., 2015. Authorised EU health claims for the essential fatty acids: n−6 linoleic (18:2, n−6) and n−3 α-linolenic (18:3, n−3) acids. Sadler, M.J. (Ed.), Foods, Nutrients and Food Ingredients with Authorised EU Health Claims, vol. 2, Woodhead Publishing, Cambridge, United Kingdom, (Chapter 12).
World Health Organization, 2008. Raised Cholesterol. WHO. Available from: http://www.who.int/gho/ncd/risk_factors/cholesterol_text/en/.
World Health Organization, 2014. Obesity Data and Statistics. WHO. Available from: http://www.euro.who.int/en/health-topics/noncommunicable-diseases/obesity/data-and-statistics.
Wuolijoki, E., Hirvelä, T., Ylitalo, P., 1999. Decrease in serum LDL cholesterol with microcrystalline chitosan. Methods Find. Exp. Clin. Pharmacol. 21 (5), 357–361.
Zacour, A., Silva, M., Cecon, P., Bambirra, E., Vieira, E., 1992. Effect of dietary chiton on cholesterol absorption and metabolism in rats. J. Nutr. Sci. Vitaminol. 38, 609–613.
Zahorska-Markiewicz, B., Krotkiewski, M., Olszanecka-Glinianowicz, M., Zurakowski, A., 2002. Effect of chitosan in complex management of obesity. Pol. Merkur. Lekarski 13 (74), 129–132.

Further Reading

Deuchi, K., Kanauchi, O., Imasato, Y., Kobayashi, E., 1995. Effect of the viscosity or deacetylation degree of chitosan on fecal fat excreted from rats fed on a high fat diet. Biosci. Biotechnol. Biochem. 59, 781–785.
EFSA Panel on Dietetic Products, Nutrition and Allergies (NDA), 2011. Guidance on the scientific requirements for health claims related to antioxidants, oxidative damage and cardiovascular health. EFSA J. 9 (12), 2474.
Lehtimaki, T., Wuolijoki, E., Hirvela, T., Lehtinen, S., Ylitalo, R., 1999. Decrease in serum LDL cholesterol with microcrystalline chitosan in healthy subjects. Atherosclerosis 144 (1), 189.
National Heart, Lung, and Blood Institute, 2005. High Blood Cholesterol: What You Need to Know. US National Heart, Lung, and Blood Institute. Available from: http://www.nhlbi.nih.gov/health/resources/heart/heart-cholesterol-hbc-what-html.
Nichols, M., Townsend, N., Luengo-Fernandez, R., Leal, J., Gray, A., Scarborough, P., Rayner, M., 2012. European Cardiovascular Disease Statistics 2012. European Heart Network, Brussels, European Society of Cardiology, Sophia Antipolis, France.
The EU Register of Nutrition and Health Claims made on foods. Available from: http://ec.europa.eu/food/safety/labelling_nutrition/claims/register.

Authorised EU health claim for meal replacements

12

Sarah Kuczora
Dudley Metropolitan Borough Council, Dudley, United Kingdom

12.1 Introduction

This chapter discusses the following authorised EU article 13.1 health claims on meal replacements for weight control:

- Substituting two of the main daily meals of an energy-restricted diet with meal replacements contributes to weight loss.
- Substituting one of the main daily meals of an energy-restricted diet with a meal replacement contributes to the maintenance of weight after weight loss.

The process of health claim authorisation is subject to Regulation (EC) No. 1924/2006 on nutrition and health claims which requires all submitted claims to receive a scientific opinion from the European Food Safety Authority (EFSA) Panel on Dietetic Products, Nutrition and Allergies (NDA) (the Panel). On receiving this Opinion, the European Commission either authorises or rejects the claim. Both authorised and non-authorised claims can be found on the EU Register of Nutrition and Health Claims (http://ec.europa.eu/nuhclaims/).

The first part of this chapter considers the characteristics that define 'meal replacements' and summarises the evidence that was submitted to support the authorisation of these health claims. Recent regulatory changes will impact the characteristics of meal replacements and these are discussed in detail. Following this, the relevance of these health claims for consumers is considered, including the burden of obesity on public health and the role of meal replacements in supporting weight loss and weight maintenance goals. The current market for meal replacements and opportunities for product development are reviewed, additionally future trends in this market are highlighted.

12.2 Characterisation of meal replacements

The food which is the subject of these health claims is 'meal replacement for weight control'. At the time the claims were authorised, their essential composition was laid out in Annex I of Directive 96/8/EC on foods intended for use in energy-restricted diets for weight reduction. This states that the energy provided by a meal replacement shall not be less than 840 kJ (200 kcal) and shall not exceed 1680 kJ/meal

Foods, Nutrients and Food Ingredients with Authorised EU Health Claims. http://dx.doi.org/10.1016/B978-0-08-100922-2.00012-7
Copyright © 2018 Elsevier Ltd. All rights reserved.

(400 kcal/meal). Additionally, conditions for the nutritional content are outlined. Meal replacements shall provide not less than 25% (and not more than 50%) of the total energy of the product as protein, not more than 30% of total available energy as fat, not less than 1 g linoleic acid (in the form of glycerides), at least 30% of the adult dietary reference values for for vitamins and minerals specified in the Annex and at least 500 mg of potassium.

12.2.1 Replacement of Directive 2009/39/EC with Regulation (EU) No. 609/2013

From 20 July 2016 Directive 2009/39/EC on foodstuffs intended for particular nutritional uses and the specific Directives adopted under this framework (including Directive 96/8/EC) were replaced by Regulation (EU) No. 609/2013 on food intended for infants and young children, food for special medical purposes and total diet replacement for weight control (Food for Specific Groups).

Regulation (EU) No. 609/2013 doesn't include meal replacements for weight control products within its scope. This is due to increasing numbers of foods appearing in the market for the general population which carry similar statements. These are presented as health claims for weight control as defined in Regulation (EC) No. 1924/2006 on nutrition and health claims made on foods. In a statement on 'the conditions of use for health claims related to meal replacements for weight control' EFSA confirmed that in the future such statements on products should be solely regulated under Regulation (EC) No. 1924/2006 and comply with the requirements set out in the Annex of Regulation (EU) No. 432/2012 that authorises the list of Article 13.1 health claims.

On 24 August 2016, Regulation (EU) 2016/1413 was published to amend Regulation (EU) No. 432/2012 with regards to meal replacements for weight control. It is stated that a transitional period *should be provided* to enable food businesses to adapt to the necessary changes. The period is not specified in the Regulation; however, a reference to the draft regulation in B.08 of the 'Summary Report of the Standing Committee on Plants, Animals, Food and Feed held in Brussels on 12 April 2016' states a 3-year transition period.

The main points for consideration are as follows:

1. *Energy content*: The 840–1680 kJ (200–400 kcal) value detailed in Directive 96/8/EC needs revising to take account of the claimed effect relating to a maximum value of 1046 kJ (250 kcal). Therefore, the energy content shall not be less than 840 kJ (200 kcal) and shall not exceed 1046 kJ (250 kcal).
2. *Macronutrient content*: The requirements set out in Directive 96/8/EC on fat, protein and amino acids should be maintained.
3. *Micronutrient content*: Required to provide 30% of the nutrient reference values (NRVs) laid down in PART A of Annex XIII of Regulation (EU) 1169/2011 (Food Information for Consumers Regulation), instead of 30% of the NRVs laid down in Annex I of Directive 96/8/EC.*
4. *Claim wording*: Slight amendment so it is clear that it relates to substituting one *main* meal.**

*The following additional conditions regarding micronutrient content are also laid out in Regulation (EU) 2016/1413:

- There is no requirement for meal replacements for weight control to provide 30% of the NRVs for fluoride, chromium, chloride and molybdenum as set out in Annex XIII of Regulation (EU) No. 1169/2011. This is due to the fact that Directive 96/8/EC did not require the addition of these micronutrients to meal replacements for weight control, and the claimed effect is related to the controlled energy content and relatively high-protein/low-fat content.
- Although Annex XIII of Regulation (EU) No. 1169/2011 does not set an NRV for sodium, the requirement to provide 30% of the sodium amount per meal as laid down in Annex I of Directive 96/8/EC should be maintained.
- Rather than providing 30% of the NRV for potassium as set out in Annex XIII of Regulation (EU) No. 1169/2011, a minimum of 500 mg potassium/meal as specified in Annex I of Directive 96/8/EC should be maintained.

**The claim wording for both claims is confirmed in the Annex to Regulation (EU) 2016/1413 as:

- Substituting one of the main daily meals of an energy-restricted diet with a meal replacement contributes to the maintenance of weight after weight loss.
- Substituting two of the main daily meals of an energy-restricted diet with meal replacements contributes to weight loss.

In a statement published on *13 November 2015*, EFSA (2015) provided advice on whether proposed compositional changes would affect the substantiation of the two health claims under discussion in this chapter. EFSA acknowledged that there would be some changes to the micronutrient content of meal replacements (Table 12.1). However, given that the mechanisms by which meal replacements exert their effects were attributed mainly to their controlled energy content and relatively high-protein/low-fat content, the Panel concluded that these changes would not affect the scientific substantiation of these health claims.

12.3 Summary of the authorised claims

In accordance with Regulation (EC) No. 1924/2006 on nutrition and health claims, the EFSA Panel was asked to provide a scientific opinion on the scientific substantiation of health claims related to meal replacements for weight control. Its opinion is based on references provided by Member States or directly from stakeholders. This section considers the evidence provided to substantiate each claim and discusses the approved wording of the claims and their conditions of use.

12.3.1 Substituting two of the main daily meals of an energy-restricted diet with meal replacements contributes to weight loss

12.3.1.1 Substantiating evidence

In total, 25 references were submitted to the Panel which included two meta-analyses of randomised controlled trials (RCT) in human subjects (Anderson et al., 2004;

Table 12.1 Nutrient reference values (NRVs) and differences in the nutrient content of meal replacements for weight control, assuming 30% of the NRVs laid down in Part A of Annex XIII of Regulation (EU) 1169/2011 instead of 30% of NRVs laid down in Annex I of Directive 96/8/EC

	NRVs	Meal replacements		
	Directive 96/8/EC	Regulation (EC) 1169/2011	Difference/ serving[a]	% Difference/ serving
Vitamin A (μg RE)[b]	700	800	+30	+14
Vitamin D (μg)	5	5	0	0
Vitamin E (mg TE)[b]	10	12	+0.6	+19
Vitamin K (μg)	—	75	—[c]	—[c]
Vitamin C (mg)	45	80	+10.5	+77
Thiamine (mg)	1.1	1.1	0	0
Riboflavin (mg)	1.6	1.4	−0.06	−13
Niacin (mg NE)[b]	18	16	−0.6	−12
Vitamin B_6 (mg)	1.5	1.4	−0.03	−7
Folate (μg)[d]	200	200	0	0
Vitamin B_{12} (μg)	1.4	2.5	+0.33	+78
Biotin (μg)	15	50	+10.5	+233
Pantothenic acid (mg)	3	6	+0.9	+100
Chloride (mg)	—	800	—[c]	—[c]
Calcium (mg)	700	800	+30	+14
Phosphorus (mg)	550	700	+45	+27
Potassium (mg)	3100	2000	—[e]	—[e]
Iron (mg)	16	14	−0.6	−13
Zinc (mg)	9.5	10	+0.15	+5
Copper (mg)	1.1	1	−0.03	−10
Iodine (μg)	130	150	+6	+15
Selenium (μg)	55	55	0	0
Sodium (mg)	575	—	—[c]	—[c]
Magnesium (mg)	150	375	+67.5	+150
Manganese (mg)	1	2	+0.4	+100
Fluoride (mg)	—	3.5	—[c]	—[c]
Chromium (μg)	—	40	—[c]	—[c]
Molybdenum (μg)	—	50	—[c]	—[c]

[a]Difference per meal replacement (i.e. per serving).

[b]The units of measurement given in Regulation 1169/2011 for vitamin A (μg), vitamin E (mg) and niacin (mg) refer to retinol equivalents (RE), tocopherol equivalents (TE) and niacin equivalents (NE) respectively, as in Directive 96/8/EC.

[c]Differences have not been calculated for nutrients for which NRVs are not given in either Directive 96/8/EC or Regulation 1169/2011.

[d]The NRVs given in Regulation 1169/2011 for 'folic acid' actually refer to folate, as in Directive 96/8/EC.

[e]A minimum of 500 mg of potassium/meal replacement (i.e. per serving) was required in Directive 96/8/EC which is below 30% of either reference value (i.e. 3100 in Directive 96/8/EC and 2000 mg in Regulation 1169/2011 respectively).

Source: Table reproduced from Table 1 in EFSA Panel on Dietetic Products, Nutrition and Allergies (NDA), 2015. Statement on the conditions of use for health claims related to meal replacements for weight control. EFSA J. 13 (11), 4287.

Heymsfield et al., 2003). The Panel considered that these meta-analyses covered most of the pertinent studies submitted in relation to the claim and so based the assessment on these.

Heymsfield et al. (2003) considers six RCTs in a meta-analysis where subjects were asked to replace either one (typically weight maintenance phase) or two (usually weight loss phase) meals with meal replacement products and to follow a standard energy-restricted diet (>800 and <1600 kcal/day). Alternatively subjects followed a standard energy-restricted diet that provided the same calorie content but did not include meal replacement products. Study durations ranged between 3 and 51 months.

The meal replacements used were commercial products that usually contained up to 250 kcal/serving and in general were compliant with the characterisation described earlier. The exception was that protein content was typically lower at 18%–25% of energy as protein, compared with the stated minimum of 25% energy previously.

Almost 500 subjects were included in the meta-analysis, of which 249 were randomised to receive meal replacements. Subjects included those who were overweight and obese, having an initial BMI between 25 and 40 kg/m^2 (average 31 kg/m^2). Weight loss was evaluated after 3 and 12 months. Dropout rates were equivalent between groups at 3 months and significantly lower in the meal replacement group at 12 months. In both groups significant weight loss was observed at each time point (1 study of 75 subjects only included data at 3 months). However, in subjects receiving meal replacements, weight loss was significantly greater (about 7%–8% of the initial body weight) than in subjects following a typical energy-restricted diet (about 3%–7% of the initial body weight). In the meal replacement group, a small proportion of subjects had type 2 diabetes. When these subjects were omitted the weight loss difference was slightly more pronounced.

A random effects meta-analysis estimate indicated a significantly greater weight loss at 3 months and 12 months, where the weight loss difference between groups was 2.54 and 2.43 kg, respectively. In its opinion the Panel notes that most of the studies evaluated data for completers only (as per their protocol), concluding that the most plausible explanation for the observed effects is better compliance with the intervention in the meal replacement group.

Anderson et al. (2004) includes 47 intervention studies as follows: 4 studies assessing the effects of meal replacements (at least 2 meal replacements/day as part of an energy-restricted diet), 6 studies using energy-restricted diets (>1500 kcal/day), 10 studies using low-energy diets (800–1500 kcal/day), 19 studies using very low-calorie diets (VLCDs) (up to 800 kcal/day) and 8 studies using soy diets (up to 800 kcal/day).

As in Heymsfield et al.'s (2003) study, all studies in the meta-analysis used commercial products for the replacement of meals. These products usually contained up to 250 kcal/serving and in general were compliant with the characterisation described previously. The exception was that the protein content provided was typically lower at 18%–25% of energy as protein, compared with the stated minimum of 25% energy earlier. The meta-analysis included 470 female subjects and 133 male subjects (603 in total). Subjects were overweight or obese, with BMI ranging from 28–34 kg/m^2. Weight loss data were reported after 24 weeks (approximately 6 months) of treatment.

Of the four studies assessing the effects of meal replacements, two were addressed by Heymsfield et al. (2003) (Ashley et al., 2001; Ditschuneit et al., 1999), whereas one was published after the end date for the literature search (Bowerman et al., 2001). These studies demonstrated an average weight loss of 9.3% for women and 8.6% for men. This weight loss was almost double that achieved with traditional energy-restricted diets having the same overall energy content. As stated by Heymsfield et al. (2003), most of the studies presented data on completers alone.

Different mechanisms have been suggested by which replacing one or two meals per day with meal replacements could promote weight loss as compared with traditional energy-restricted diets. It is noted by Anderson et al. (2004) that meal replacements seem to increase compliance with energy-restricted diets. This could be partially due to that fact that they offer easy and 'ready-to-eat' meals with a controlled energy content, and partially because the high-protein, low-fat composition promotes sustained satiety (the feeling of fullness after a meal that suppresses the desire to eat). Alternatively, Krieger et al. (2006) and Westerterp-Platenga et al. (2009) suggest the macronutrient composition of the meal replacements may result in increased energy expenditure and sparing of lean body mass which could explain their effects on weight loss and maintenance.

Based on these two meta-analyses, the Panel concluded that a cause-and-effect relationship has been established between the consumption of meal replacements (instead of regular meals), while following an energy-restricted diet, and reduction in body weight.

12.3.1.2 Claim wording

The Panel stated that the following wording reflects the scientific evidence presented: 'Substituting two daily meals with meal replacements in the context of energy restricted diets helps to lose weight'. However, the final wording agreed during the authorisation procedures is 'Substituting two daily meals of an energy restricted diet with meal replacements contributes to weight loss'. Following publication of Regulation (EU) 2016/1413 the wording is now agreed to be 'Substituting two of the main daily meals of an energy restricted diet with a meal replacement contributes to weight loss'.

12.3.1.3 Conditions and restrictions of use

To bear the claim, a food should comply with specifications laid down in Regulation (EU) 2016/1413. Information shall be provided to the consumer on the importance of maintaining an adequate daily fluid intake and on the fact that the products are useful for the intended use only as part of an energy-restricted diet and that other foodstuffs should be a necessary part of such a diet. To achieve the claimed effect, two of the main daily meals should be substituted with meal replacements daily. The target population is overweight people in the general population who want to reduce their body weight. In the context of the claim, weight loss without the achievement of a normal body weight is considered to be beneficial to health. The claimed effect is 'safe, effective and healthy weight loss; greater weight loss than with conventional calorie counting'.

12.3.2 Substituting one of the main daily meals of an energy-restricted diet with a meal replacement contributes to the maintenance of weight after weight loss

12.3.2.1 Substantiating evidence

In total, 14 references were submitted for substantiation of the claimed effect; however, a number of these were not considered pertinent to the claim. The reasons provided were short follow-up (Anderson et al., 2004; Heber et al., 1994; Winick et al., 2002); data based on pharmacologically treated type 2 diabetic subjects which aren't representative of the target population (Redmon et al., 2003; Yip et al., 2001) and original data not reported (e.g. duplicate, letters and comments) or were limited to abstracts with a lack of sufficient information to carry out a full scientific evaluation.

The Panel considered the studies discussed by Heymsfield et al. (2003) which provided data on weight loss achieved from 3 to 12 months, were relevant. This was due to the greatest weight loss occurring in the first 3 months and that, after 3 months, any further weight loss was considered to substantiate a claim on weight maintenance after weight loss. During this time one meal and one snack are typically substituted with meal replacements.

Three additional references reporting longer-term follow-ups of some studies from the meta-analysis by Heymsfield et al. (2003) were also considered pertinent. These studies reported sustained weight loss after 27 months (Ditschuneit et al., 1999) which was maintained after 4 years (−8.4% of initial body weight) (Flechtner-Mors et al., 2000), with one meal and one snack replacement. Rothacker (2000) also demonstrated weight loss in 84 overweight women (−4.2 kg) and 50 overweight males (−5.8 kg) after a 5-year self-managed meal replacement weight control programme, compared with matched controls who gained 6.5 and 6.7 kg respectively during the same period of time.

Most of the studies demonstrated significantly better weight maintenance after weight loss when meal replacements were included in the maintenance phase. As in the studies presented for weight loss, all meals were commercially sourced and generally compliant with the composition in Section 12.2. The proposed mechanisms by which meal replacements could exert the claimed effect are generally in relation to their controlled energy content and relatively high-protein, low-fat macronutrient content.

The Panel concluded that a cause-and-effect relationship has been established between the consumption of meal replacements in substitution for regular meals and maintenance of body weight after weight loss.

12.3.2.2 Claim wording

The Panel stated that the following wording reflects the scientific evidence presented: 'Substituting one or two daily meals with meal replacements helps to maintain body weight after weight loss'. However, the final wording agreed during the authorisation procedures is 'Substituting one daily meal of an energy restricted diet with a meal replacement contributes to the maintenance of weight after weight loss'. Following publication of Regulation (EU) 2016/1413 the wording is now agreed to be 'Substituting one of the main daily meals of an energy restricted diet with a meal replacement contributes to the maintenance of weight after weight loss'.

12.3.2.3 Conditions and restrictions of use

To bear the claim, a food should comply with specifications laid down in Regulation (EU) 2016/1413. Information shall be provided to the consumer on the importance of maintaining an adequate daily fluid intake and on the fact that the products are useful for the intended use only as part of an energy-restricted diet and that other foodstuffs should be a necessary part of such a diet. To achieve the claimed effect, one main meal should be substituted with one meal replacement daily. The target population is overweight people in the general population who want to maintain their body weight after significant weight loss. In the context of the claim, the maintenance of weight loss in overweight subjects without achieving a normal body weight is considered beneficial to health.

12.4 Other relevant legislation

12.4.1 On-hold claim for very low-calorie diets (VLCDs) and reduction in body weight

In 2011 EFSA gave a positive opinion on the claim that VLCDs lead to a reduction in body weight, where the claimed effect was 'safe and effective weight loss, long term weight maintenance' (EFSA Panel, 2011). However, this claim has been on-hold since 2012, with the European Commission neither authorising nor refusing authorising the claim to date. As mentioned earlier in Section 12.2.1, total diet replacements for weight control are being regulated under Regulation (EU) No. 609/2013. Article 11 of the Regulation empowers the European Commission to adopt delegated acts for each food within its scope (including total diet replacements for weight control). The Delegated Act was published on 2nd June 2017 and states (Article 6) that nutrition and health claims are not allowed on such products, except for a nutrition claim for 'added fibre' which can be used provided the product contains not less than 10g fibre. Therefore the health claim for VLCD is expected to be refused authorisation, although at the time of writing no official refusal has been made.

12.5 Consumer issues

12.5.1 The need for meal replacement products

The prevalence of overweight and obesity in adults is a major public health concern. Excess weight is a key risk factor for a number of chronic diseases and associated conditions, including high blood pressure, cardiovascular disease, type 2 diabetes, arthritis, gallstones, some cancers (e.g. breast and prostate), snoring and sleep apnoea, stress, anxiety and depression, and infertility.

Body mass index (BMI) is the measure commonly used to classify overweight and obesity in adults. BMI is calculated as follows:

Person's weight in kilograms divided by the square of their height in metres (kg/m²)

The World Health Organization (WHO) defines 'overweight' as a BMI ≥25 and 'obesity' as a BMI ≥30. According to WHO around 39% of the world's adult population is overweight and 13% is obese. For the European Union, the Organisation for Economic Co-operation and Development (OECD) reports that on average 52% of the adult population is either overweight or obese, whereas 17% is obese. Obesity rates vary considerably between Member States, from 8% in Romania to over 25% in the United Kingdom and Hungary. Average obesity rates are reportedly similar among men and women.

Given the high rates of overweight and obesity, and the significant health implications, it is important that individuals are supported to manage their weight. However, obesity is a complex condition. Prolonged weight gain results from a long-term energy imbalance, where energy intake from food and drink regularly exceeds the energy utilised by metabolic processes and physical activity. Physiological responses usually regulate energy balance within the body, monitoring energy intake expenditure and sending signals to control appetite. However, external societal and environmental factors have a powerful influence and can overwhelm the body's natural physiological responses, resulting in weight gain. Furthermore, an individual's susceptibility to these external influences is affected by genetic and biological factors such as age, gender and hormonal activity. The Foresight Obesity System Map attempts to demonstrate the complexity of obesity and the relationship between the wide range of factors that influence it. The full map can be found at https://www.gov.uk/government/uploads/system/uploads/attachment_data/file/296290/obesity-map-full-hi-res.pdf.

Food intake is one of the key behavioural factors influencing weight, and healthy weight loss should involve long-term lifestyle changes, including changes in eating habits. It is reported that even modest weight loss (e.g. 5%–10% of body weight) can help to reduce the risk of developing chronic diseases. Therefore, the meal replacement products described in this chapter are likely to help overweight or obese consumers achieve weight loss that has a positive effect on their health. For example the evidence supporting the claim on weight loss demonstrated that the use of meal replacements within an energy-restricted diet led to average losses of initial body weight between 7% and 8% (Heymsfield et al., 2003) and 9.3% for women and 8.6% for men (Anderson et al., 2004). However, for many consumers it can be difficult to sustain weight loss in the long term. Therefore, products which help them to maintain their weight after weight loss will also be beneficial as part of a longer-term lifestyle change. The use of meal replacements has been shown to support the maintenance of weight loss for up to 4 and 5 years [Flechtner-Mors et al. (2000) and Rothacker (2000) respectively].

12.5.2 Consumer understanding and flexible use of claim wording

Meal replacement products have been around for some time and therefore it is likely that consumer understanding of their role in weight loss and weight maintenance is high. A number of meal replacement products currently carry one of the approved claims for weight loss or weight maintenance. Some products also carry other authorised health and nutrition claims. Examples of claims used on various meal replacement products are provided further.

Flexibility of claim wording is permitted by the Regulation, providing that the wording of the claim has the same meaning for consumers as the authorised claim. Importantly, the claim must not be made 'stronger' than the authorised claim. For example 'contributes to' could be replaced by 'plays a role' or 'supports', but not by 'optimises' or 'stimulates'. In general, it is advised that food business operators stick as closely as possible to the authorised wording. This will ensure that consumers are provided with the right information and help enforcement officers to assess whether claims are compliant with the law. It should also be noted that the EU Regulation (EC) No. 1924/2006 on Nutrition and Health Claims prohibits any reference to the rate or amount of weight loss that may result from the use of any food product. Therefore, such claims should not be used in the labelling, presentation and advertising of products to consumers.

New products enter the market regularly, while others are delisted; however, examples of meal replacement products recently available in the United Kingdom market, and their advertised nutrition and/or health claims, can be found in Table 12.2. Whilst this represents a good range of available meal replacement products in the UK market, this list is not exhaustive and is only used to give an indication of the types of products available and their nutrition and health claims.

As can be seen in Table 12.2, some manufacturers have used flexible wording to present the claims, for example 'promotes effective weight control', 'supports healthy weight loss', and 'safe, healthy and effective weight loss'. These statements typically appear on the front-of-pack image provided and are often followed by an indication [e.g. an asterisk (*)] that further information is provided elsewhere on the pack. This is likely to include the authorised claim wording (although the back-of-pack information is not available online) as well as other required information (e.g. a statement indicating the importance of a varied and balanced diet and a healthy lifestyle). Other manufacturers have simply used the approved claim wording 'substituting two daily meals of an energy restricted diet with meal replacements contributes to weight loss', either on front of pack or within the product description on the website; this practice would be advisable for all products to prevent any confusion. For the future, food manufacturers should note the change in claim wording following publication of Regulation (EU) 2016/1413 as described earlier in the chapter.

In some cases products are described as a 'meal replacement for weight control' or bear the statement 'can help as part of a calorie controlled diet'. Labelling of products in this way was required by Directive 2009/39/EC on foodstuffs intended for particular nutritional uses. However, this legislation has now been replaced by Regulation (EU) No. 609/2013, and these statements are now subject to the Nutrition and Health Claims Regulation. If still used, food manufacturers should accompany the statements with the full claim wording, either on pack or in the website description where sold online. This requirement is outlined in B.08 of the 'Summary Report of the Standing Committee on Plants, Animals, Food and Feed held in Brussels on 12 April 2016'.

Table 12.2 Examples of meal replacement products currently available in the UK market and their advertised nutrition and/or health claims

Brands	Products	Advertised nutrition and/or health claims	Date information was last updated on
Almased	Almased Meal Replacement Soya Honey and Yoghurt (500 g)	Promotes effective weight control	23/12/2015
Atkins	Atkins Cuisine Penne (250 g)	High in protein High in fibre Carbohydrate reduced	24/09/2014
	Atkins Advantage Chocolate Shake Mix (10 × 37 g)	Low in sugar High in protein High in fibre Can help as part of a calorie-controlled diet	21/07/2014
	Atkins Advantage Cappuccino Shake (6 × 330 mL)	High protein Low sugar No added sugar	22/07/2014
Boots	Boots Slender Shake Chocolate Flavour Milkshake Powder (360 g)	Meal replacement for weight control	13/10/2014
	Boots Slender Shake Banana Flavour Milkshake Powder (360 g)	Meal replacement for weight control	13/10/2014
	Boots Slender Shake Vanilla Flavour Milkshake Powder (360 g)	Meal replacement for weight control	13/10/2014
	Boots Slender Shake Strawberry Flavour Milkshake Powder (360 g)	Meal replacement for weight control	13/10/2014
Celebrity Slim	Celebrity Slim ACTIVE Caramel Crunch Meal Replacement Bar (58 g)	Supports healthy weight loss High protein	25/03/2015
	Celebrity Slim ACTIVE Caramel Crunch Meal Replacement Bar (12 × 58 g)	Supports healthy weight loss High protein	25/03/2015

(*Continued*)

Table 12.2 Examples of meal replacement products currently available in the UK market and their advertised nutrition and/or health claims (*cont.*)

Brands	Products	Advertised nutrition and/or health claims	Date information was last updated on
	Celebrity Slim Double Chocolate Meal Replacement Bar (58 g)	Meal replacement for weight control	25/03/2015
	Celebrity Slim Double Chocolate Meal Replacement Bar (12×)	Meal replacement for weight control	25/03/2015
	Celebrity Slim Fruit and Nut Meal Bar (55 g)	Can help as part of a calorie-controlled diet	14/08/2014
	Celebrity Slim Fruit and Nut Meal Bar (12×)	Can help as part of a calorie-controlled diet	14/08/2014
	Celebrity Slim 7 Day Assorted Shake Pack	Can help as part of a calorie-controlled diet	14/08/2014
	Celebrity Slim Banana 7 Day Shake Pack	Safe, healthy and effective weight loss: meal replacement for weight control Calorie controlled and high in protein	14/08/2014
	Celebrity Slim Chocolate 7 Day Shake Pack	Safe, healthy and effective weight loss: meal replacement for weight control Calorie controlled and high in protein	16/11/2014
	Celebrity Slim Strawberry 7 Day Shake Pack	Safe, healthy and effective weight loss: meal replacement for weight control Calorie controlled and high in protein	24/07/2014
	Celebrity Slim Chocolate Handy Shake Pack (7 × 55 g)	Safe, healthy and effective weight loss: meal replacement for weight control	24/03/2015
	Celebrity Slim Strawberry Handy Shake Pack (7 × 55 g)	Safe, healthy and effective weight loss: meal replacement for weight control	24/03/2015
	Celebrity Slim Active Shake Chocolate With Sweetener (21 × 40 g)	Supports healthy weight loss: meal replacement for weight control High protein	25/07/2014
	Celebrity Slim Active Shake Strawberry (21 × 40 g)	Supports healthy weight loss: meal replacement for weight control High protein	05/06/2015

	Celebrity Slim Assorted Handy Soup Pack (7 × 55 g)	Meal replacement for weight control	24/03/2015
	Celebrity Slim Banana Single Sachet (55 g)	Meal replacement for weight control	06/11/2014
	Celebrity Slim Vanilla Single Sachet Shake (55 g)	Meal replacement for weight control	06/11/2014
	Celebrity Slim Strawberry Sachet (55 g)	Meal replacement for weight control	14/08/2014
	Celebrity Slim Café Latte Single Sachet (55 g)	Meal replacement for weight control	07/11/2014
	Celebrity Slim Cookies and Cream Single Sachet (55 g)	Meal replacement for weight control	07/11/2014
	Celebrity Slim Variety Shake Pack (20 × 55 g)	Supports healthy weight loss: meal replacement for weight control High protein	24/03/2015
Complan	Complan Milkshake Chocolate Flavour (250 mL)	Milk drink with extra nutrients: high in calcium and vitamin D	23/07/2014
	Complan Milkshake Strawberry Flavour (250 mL)	Milk drink with extra nutrients: high in calcium and vitamin D	15/09/2014
	Complan Chocolate Flavour Nutritional Drink (4 × 55 g)	B vitamins, including vitamin B_6 and B_{12} for the reduction of tiredness and fatigue Vitamin C and iron which help to support normal immune function Protein which contributes to muscle growth and maintenance Calcium and vitamin D needed for maintenance of normal bones	07/07/2015
	Complan Strawberry Flavour Nutritional Drink (4 × 55 g)	B vitamins, including vitamin B_6 and B_{12} for the reduction of tiredness and fatigue Vitamin C and iron which help to support normal immune function Protein which contributes to muscle growth and maintenance Calcium and vitamin D needed for maintenance of normal bones	07/07/2015

(*Continued*)

Table 12.2 Examples of meal replacement products currently available in the UK market and their advertised nutrition and/or health claims (*cont.*)

Brands	Products	Advertised nutrition and/or health claims	Date information was last updated on
	Complan Tasty Chicken Flavour Soup (4 × 55 g)	B vitamins, including vitamin B_6 and B_{12} for the reduction of tiredness and fatigue Vitamin C and iron which help to support normal immune function Protein which contributes to muscle growth and maintenance Calcium and vitamin D needed for maintenance of normal bones	07/07/2015
	Complan Original Flavour (425 g)	B vitamins, including vitamin B_6 and B_{12} for the reduction of tiredness and fatigue Vitamin C and iron which help to support normal immune function Protein which contributes to muscle growth and maintenance Calcium and vitamin D needed for maintenance of normal bones	07/07/2015
Forza	Forza 14 Day Starter Pack	Meal replacement for weight control	23/12/2015
	Forza 7 Day Starter Pack	Meal replacement for weight control	23/12/2015
	Forza Hydratim Weightloss Capsules (3-month supply)	Glucomannan can delay gastric emptying and induce satiety, leading to a decrease in subsequent food intake	23/12/2015
	Forza One Month Supply	Meal replacement for weight control	23/12/2015
	Forza Shake it Slim Meal Replacement Refill Pack	Meal replacement for weight control	23/12/2015
	Forza Shake it Slim Meal Replacement Starter Pack (7-day supply)	Meal replacement for weight control	Not given

MaxiNutrition	MaxiNutrition Sculptress High Protein Diet Meal Replacement Bar Caramel Crunch (60 g)	Supports weight loss Zinc for health skin, hair and nails 17-g protein helps support healthy muscle tone B vitamins for energy release	25/09/2014
	MaxiNutrition Sculptress High Protein Diet Meal Replacement Milk Strawberry Flavour (330 mL)	Supports weight loss Zinc for health skin, hair and nails 23-g protein helps support healthy muscle tone B vitamins for energy release	02/12/2014
	MaxiNutrition Sculptress High Protein Diet Meal Replacement Shake Strawberry Sundae Flavour (700 g)	Supports weight loss Zinc for health skin, hair and nails 17-g protein helps support healthy muscle tone B vitamins for energy release	23/09/2014
	MaxiNutrition Sculptress High Protein Diet Meal Replacement Shake Chocolate Delight Flavour (700 g)	Supports weight loss Zinc for health skin, hair and nails 17-g protein helps support healthy muscle tone B vitamins for energy release	26/08/2014
OatWell	OatWell Strawberry Shake Sachets (14 × 15 g)	Oat beta-glucan to reduce cholesterol High in fibre	13/08/2014
PhD	PhD Woman Meal Replacement Chocolate Peanut Flavour Bar (60 g)	Meal replacement bar for weight control	24/04/2015
	PhD Woman Meal Replacement Chocolate Cookie (770 g)	Supports healthy weight loss Vitamin B_{12} helps reduce tiredness and fatigue	30/09/2014
	PhD Woman Meal Replacement Vanilla Crème (770 g)	Supports healthy weight loss Vitamin B_{12} helps reduce tiredness and fatigue Helps maintain weight loss Helps maintain muscle tone	30/09/2014

(Continued)

Table 12.2 Examples of meal replacement products currently available in the UK market and their advertised nutrition and/or health claims (*cont.*)

Brands	Products	Advertised nutrition and/or health claims	Date information was last updated on
	PhD Woman Meal Replacement Strawberry Delight With Sweetener (770 g)	Supports healthy weight loss Vitamin B_{12} helps reduce tiredness and fatigue	30/09/2014
	PhD Woman Caramel Crunch Meal Replacement Bar (60 g)	Meal replacement bar for weight control	27/08/2015
Slim-Fast	Slim-Fast Café Latte Flavour Shakes (6 × 325 mL)	Healthy effective weight loss High protein Source of fibre	18/09/2015
	Slim-Fast Café Latte Milk Shake (325 mL)	Substituting two daily meals of an energy-restricted diet with meal replacements contributes to weight loss High protein High fibre	24/08/2014
	Slim-Fast Chunky Chocolate Flavour Shakes (6 × 325 mL)	Healthy effective weight loss High protein Source of fibre	18/09/2015
	Slim-Fast Chunky Chocolate Milk Shake (325 mL)	Substituting two daily meals of an energy-restricted diet with meal replacements contributes to weight loss High protein High fibre	24/08/2015
	Slim-Fast Summer Strawberry Flavour Shakes (6 × 325 mL)	Healthy effective weight loss High protein Source of fibre	18/09/2015

	Slim-Fast Summer Strawberry Milk Shake (325 mL)	Substituting two daily meals of an energy-restricted diet with meal replacements contributes to weight loss High protein Source of fibre	24/08/2015
	Slim-Fast Simply Vanilla Milk Shake (325 mL)	Substituting two daily meals of an energy-restricted diet with meal replacements contributes to weight loss High protein High fibre	—
	Slim-Fast Blissful Banana Milk Shake (325 mL)	Substituting two daily meals of an energy-restricted diet with meal replacements contributes to weight loss High protein High fibre	—
	Slim-Fast Choca Mocha Milk Shake (325 mL)	Substituting two daily meals of an energy-restricted diet with meal replacements contributes to weight loss High protein High fibre	—
	Slim-Fast Chicken Tikka Masala Noodle Box (250 g)	Healthy effective weight loss Low fat High protein and fibre	18/08/2015
	Slim-Fast Spaghetti Bolognese Noodle Box (250 g)	Healthy effective weight loss Low fat High protein and fibre	18/09/2015

(Continued)

Table 12.2 Examples of meal replacement products currently available in the UK market and their advertised nutrition and/or health claims (*cont.*)

Brands	Products	Advertised nutrition and/or health claims	Date information was last updated on
	Slim-Fast Spicy Thai Noodles (240 g)	Healthy effective weight loss Low fat Source of protein High fibre	18/09/2015
	Slim-Fast Blissful Banana Shake (438 g)	Substituting two daily meals of an energy-restricted diet with meal replacements contributes to weight loss High protein Source of fibre	24/08/2015
	Slim-Fast High-Protein Chunky Chocolate Shake (450 g)	Substituting two daily meals of an energy-restricted diet with meal replacements contributes to weight loss High protein Source of fibre	24/08/2015
	Slim-Fast Summer Strawberry Shake (438 g)	Substituting two daily meals of an energy-restricted diet with meal replacements contributes to weight loss High protein Source of fibre	08/12/2015
	Slim-Fast Simply Vanilla Shake (438 g)	Substituting two daily meals of an energy-restricted diet with meal replacements contributes to weight loss High protein Source of fibre	24/08/2015

Source: Data was taken from 'meal replacement' options available online at The Boots Company PLC on 30 March 2016. Available from: http://www.boots.com/en/Pharmacy-Health/Health-shop/Diet-weight-loss/Meal-replacements/All-Meal-Replacements/

Of the products surveyed, only one (PhD Woman Meal Replacement Vanilla Crème) highlighted the claim 'helps maintain weight loss', perhaps suggesting that UK consumers prefer using meal replacement products for weight loss, rather than weight maintenance. Additionally a small number of products focussed on other health and nutrition claims (e.g. related to the protein, fibre and micronutrient content), rather than the role of meal replacements in weight loss. The decision to use other claims will depend on the target consumer, some of whom may be driven by nutritional content rather than weight loss claims.

12.6 Potential of the claims in product development

As mentioned in Section 12.5.2 and outlined in Table 12.2, a wide variety of meal replacement products and brands are already available in the market. A recent report (Euromonitor, 2015) states that within the UK market the Unilever Slim Fast brand is the market leader, holding a 26% share of the market value in 2015. Typical product formats for meal replacements include soups, bars, shakes and smoothies. More recently pasta and noodle ready meals have also been added to the product portfolio of some brands which may appeal to consumers looking for more 'normal' meals during weight loss.

Shake and smoothie ranges are dominated by traditional flavours, such as strawberry, chocolate, vanilla and banana. However, some brands have developed more creative flavours, including cookies and cream, café latte and cappuccino. Meal replacement bars appear to favour chocolate, caramel and fruit-and-nut varieties. The compositional requirements for meal replacements are very specific (as defined in Section 12.2); therefore it is important that the development of meal replacement products is compliant with these criteria.

12.7 Future trends

Whilst it is clear that meal replacements are successful in supporting weight loss and weight maintenance, a recent report from Euromonitor (2015) suggests that weight management products may be viewed as 'old fashioned' or 'unreliable' as a way to lose weight. It suggests that people are turning instead to new trends, including green drinks or smoothies for breakfast, spiraliser food preparation (e.g. producing vegetable spaghetti) and wearable electronics and apps that monitor both food intake and physical activity levels. Furthermore, it is suggested that meal replacements are facing competition from fresh foods, with many consumers preferring to switch from convenient packaged foods to fresh ingredients (Euromonitor, 2016). Additionally other dietary supplements, in particular protein supplements, are increasingly popular with consumers adding protein powder to their food and drinks, as well as buying high-protein bars and ready-to-drink products. If these trends continue, weight management product brand leaders may be tempted to diversify their product ranges.

However, it is likely that meal replacement products will remain popular with those looking for a scientifically proven method of weight loss.

12.8 Conclusions

Overweight and obesity are a major public health concern in Europe. These conditions are associated with increased risk of chronic diseases such as heart disease and type 2 diabetes which reduces the quality of life, as well as placing a burden on health services. Therefore, there is a need for effective weight loss strategies that are also sustainable in the long term. Meal replacements for weight control have been shown to be effective in supporting weight loss and the maintenance of weight after weight loss, and the use of the authorised claims is common on these products. The meal replacement market appears to be facing competition from new food trends, as well as apps that track food intake and encourage increased levels of physical activity. However, with the strength of approved health claims behind them, it's likely that meal replacements are here to stay.

From 20 July 2016, Regulation (EU) No. 609/2013 on food intended for infants and young children, food for special medical purposes and total diet replacement for weight control (Food for Specific Groups) replaced Directive 2009/39/EC on foodstuffs intended for particular nutritional uses, and the specific directives adopted under this framework (including Directive 96/8/EC). This has resulted in some changes for the health claims for meal replacements which are presented in Regulation (EU) 2016/1413, and for which there is a transition period for implementation. These include a maximum energy content of 1046 kJ (250 kcal) (although this is still within the range originally specified), changes to micronutrient composition and a change in claim wording to make it clear that the claims relate to substituting one or two *main* daily meals. The EFSA Panel has concluded that such changes will not impact the scientific substantiation of the claims, as the mechanisms by which meal replacements exert their effects were attributed mainly to their controlled energy content and relatively high-protein/low-fat content. These key compositional factors will be retained following any changes.

References

Anderson, J.W., Luan, J., Hoie, L.H., 2004. Structured weight-loss programs: meta-analysis of weight loss at 24 weeks and assessment of effects of intervention intensity. Adv. Ther. 21, 61–75.

Ashley, J.M., St. Jeor, S.T., Schrage, J.P., Perumean-Chaney, S.E., Gilbertson, M.C., McCall, N.L., Bovee, V, 2001. Weight control in the physician's office. Arch. Intern. Med. 161, 1599–1604.

Bowerman, S., Bellman, M., Saltsman, P., Garvey, D., Pimstone, K., Skootsky, S., Wang, H.J., Elashoff, R., Heber, D., 2001. Implementation of a primary care physician network obesity management program. Obes. Res. 9 (Suppl. 4), 321S–325S.

Directive 2009/39/EC of the European Parliament and of the Council of 6 May 2009 on foodstuffs intended for particular nutritional uses. Off J Eur Union. L 124/21.
Ditschuneit, H.H., Flechtner-Mors, M., Johnson, T.D., Adler, G., 1999. Metabolic and weight-loss effects of a long-term dietary intervention in obese patients. Am. J. Clin. Nutr. 69, 198–204.
EFSA Panel on Dietetic Products, Nutrition and Allergies (NDA), 2011. Scientific Opinion on the substantiation of health claims related to very low calorie diets (VLCDs) and reduction in body weight (ID 1410), reduction in the sense of hunger (ID 1411), reduction in body fat mass while maintaining lean body mass (ID 1412), reduction of post-prandial glycaemic responses (ID 1414), and maintenance of normal blood lipid profile (1421) pursuant to Article 13(1) of Regulation (EC) No. 1924/2006. EFSA J. 9 (6), 2271.
EFSA Panel on Dietetic Products, Nutrition and Allergies (NDA), 2015. Statement on the conditions of use for health claims related to meal replacements for weight control. EFSA J. 13 (11), 4287.
EU, 2011. Regulation (EU) No 1169/2011 of the European Parliament and of the Council of 25 October 2011 on the provision of food information to consumers, amending Regulations (EC) No 1924/2006 and (EC) No 1925/2006 of the European Parliament and of the Council, and repealing Commission Directive 87/250/EEC, Council Directive 90/496/EEC, Commission Directive 1999/10/EC, Directive 2000/13/EC of the European Parliament and of the Council, Commission Directives 2002/67/EC and 2008/5/EC and Commission Regulation (EC) No 608/2004. Off. J. Eur. Union 22.11.2011, L 304/18.
Euromonitor, 2015. Weight management in the United Kingdom. Available from: http://www.euromonitor.com/weight-management-in-the-united-kingdom/report.
Euromonitor, 2016. Meal replacement trends: the transformation from a weight management solution to a wellbeing inspiration. Available from: http://blog.euromonitor.com/2016/01/meal-replacement-trends-the-transformation-from-a-weight-management-solution-to-a-wellbeing-inspiration.html.
Flechtner-Mors, M., Ditschuneit, H.H., Johnson, T.D., Suchard, M.A., Adler, G., 2000. Metabolic and weight loss effects of long-term dietary intervention in obese patients: four-year results. Obes. Res. 8, 399–402.
Heber, D., Ashley, J.M., Wang, H.J., Elashoff, R.M., 1994. Clinical evaluation of a minimal intervention meal replacement regimen for weight reduction. J. Am. Coll. Nutr. 13, 608–614.
Heymsfield, S.B., van Mierlo, C.A., van der Knapp, H.C., Heo, M., Frier, H.I., 2003. Weight management using a meal replacement strategy: meta and pooling analysis from six studies. Int. J. Obes. Relat. Metab. Disord. 27, 537–549.
Krieger, J.W., Sitren, H.S., Daniels, M.J., Langkamp-Henken, B., 2006. Effects of variation in protein and carbohydrate intake on body mass and composition during energy restriction: a meta-regression. Am. J. Clin. Nutr. 83, 260–274.
The Commission of the European Communities, 1996. Commission Directive 96/8/EC of 26 February 1996 on foods intended for use in energy-restricted diets for weight reduction. Off. J. Eur. Commun. L 55/22.
The European Commission, 2012. Commission Regulation (EU) No. 432/2012 of 16 May 2012 establishing a list of permitted health claims made of foods, other than those referring to the reduction of disease risk and to children's development and health. Off. J. Eur. Commun. L 136.
The European Commission, 2016. Commission Regulation (EU) 2016/1413 of 24 August 2016 amending Regulation (EU) No. 432/2012 establishing a list of permitted health claims made on foods other than those referring to the reduction of disease risk and to children's development and health. Off. J. Eur. Commun. L 230/8.

The European Parliament and the Council of the European Union, 2006. Commission Regulation (EC) No. 1924/2006 of the European Parliament and of the Council of 20 December 2006 on nutrition and health claims made on foods. Off. J. Eur. Commun. L 404.
The European Parliament and the Council of the European Union, 2013. Commission Regulation (EU) No. 609/2013 of the European Parliament and of the Council of 12 June 2013 on food intended for infants and young children, food for special medical purposes, and total diet replacement for weight control repealing Council Directive 92/52/EEC, Commission Directives 96/8/EC, 1999/21/EC, 2006/125/EC and 2006/141/EC, Directive 2009/39/EC of the European Parliament and of the Council and Commission Regulations (EC) No. 41/2009 and (EC) No. 953/2009. Off. J. Eur. Commun. L 181/35.
Redmon, J.B., Raatz, S.K., Reck, K.P., Swanson, J.E., Kwong, C.A., Fan, Q., Thomas, W., Bantle, J.P., 2003. One-year outcome of a combination of weight loss therapies for subjects with type 2 diabetes: a randomized trial. Diabetes Care 26, 2505–2511.
Rothacker, D.Q., 2000. Five-year self-management of weight using meal replacements: comparison with matched controls in rural Wisconsin. Nutrition 16, 344–348.
Westerterp-Platenga, M.S., Nieuwenhuizen, A., Tomé, D., Soenen, S., Westerterp, K.R., 2009. Dietary protein, weight loss, and weight maintenance. Annu. Rev. Nutr. 29, 21–41.
Winick, C., Rothacker, D.Q., Norman, R.L., 2002. Four worksite weight loss programs with high-stress occupations using a meal replacement product. J. Occup. Med. 52, 25–30.
Yip, I., Go, V.L., DeShields, S., Saltsman, P., Bellman, M., Thames, G., Murray, S., Wang, H.J., Elashoff, R., Heber, D., 2001. Liquid meal replacements and glycemic control in obese type 2 diabetes patients. Obes. Res. 9 (Suppl. 4), 341S–347S.

Further Reading

Commission Working Group on Claims, 2015. Working document on the technical adaptations to the Annex of Commission Regulation (EU) No. 432/2012 with respect to health claims on meal replacement for weight control. Supporting Document for the meeting of the Commission Working Group on Claims on 22 June 2015.
Commission Delegated Regulation of 2.6.2017 supplementing Regulation (EU) No 609/2013 of the European Parliament and of the Council as regards the specific compositional and information.
Department of Health, 2015. Bulletin intended for interested parties. Update from the European Commission's Working Group meeting on health claims, 9 November 2015.
EFSA Panel on Dietetic Products, Nutrition and Allergies (NDA), 2010. Scientific Opinion on the substantiation of health claims related to meal replacements for weight control (as defined in Directive 96/8/EC on energy restricted diets for weight loss) and reduction in body weight (ID 1417) and maintenance of body weight after weight loss (ID 1418) pursuant to Article 13(1) of Regulation (EC) No. 1924/2006. EFSA J. 8 (2), 1466.
The Commission of the European Communities, 2009. Commission Regulation (EC) No. 953/2009 of 13 October 2009 on substances that may be added for specific nutritional purposes in foods for particular nutritional uses. Off. J. Eur. Commun. L 269.
The European Parliament and the Council of the European Union, 2006. Commission Regulation (EC) No. 1925/2006 of the European Parliament and of the Council of 20 December 2006 on the addition of vitamins and mineral and of certain other substances to foods. Off. J. Eur. Commun. L 404.

Authorised EU health claim for arabinoxylan

13

Nicole J. Kellow, Karen Z. Walker
Monash University, Melbourne, VIC, Australia

13.1 Introduction

Dietary fibre can be broadly defined as those carbohydrate polymers and oligomers (plus lignin), which escape digestion in the small intestine, passing into the large bowel where they are partially (insoluble dietary fibre) or completely (soluble dietary fibre) fermented and broken down by the gut microbiota (Jones, 2013). Recognised health benefits associated with dietary fibre consumption include promotion of laxation and reduced gastrointestinal transit time, moderation of circulating blood lipids and attenuation of post-prandial serum glucose and insulin responses (Lafiandra et al., 2014). The non-starch polysaccharides (NSPs) which form important structural components of plant cell walls constitute a significant proportion of human dietary fibre intake. Arabinoxylans (otherwise known as pentosans) from cereal grain cell walls comprise one category of NSPs (Saeed et al., 2011). Arabinoxylans are of particular interest, as arabinoxylan-enriched foods have been reported to induce positive effects on colonic health, lipid metabolism and glycaemic control.

Inadequate consumption of dietary fibre is a major concern in western societies, with an estimated 90% of adults failing to consume the recommended total of 25–35 g of fibre/day (Reicks et al., 2014). Chronic low intake of dietary fibre has been associated with a number of deleterious health consequences including increased risk for colorectal cancer, diverticular disease, cardiovascular disease (CVD), obesity, the metabolic syndrome, pre-diabetes, type 2 diabetes and poorly controlled diabetes (Aldoori et al., 1998; Fuchs et al., 1999; Fung et al., 2002; Liu et al., 2000; McKeown et al., 2004; Pereira et al., 2004).

The world-wide prevalence of diabetes in 2013 was estimated at 382 million people and is expected to rise to 592 million by 2035 (Guariguata et al., 2014). Approximately 175 million of all diabetes cases in adults are estimated to be undiagnosed. Moreover, around 25% of the world's adults have the metabolic syndrome, which incurs a fivefold greater risk of developing type 2 diabetes (Beagley et al., 2014).

Many individuals with moderate-to-low total energy intakes due to advanced age or low levels of physical activity find it difficult to obtain an adequate dietary fibre intake. These people may benefit from regular consumption of fibre-enriched or fibre-fortified foods to achieve the daily intake of dietary fibre recommended for good health. There is an ongoing need to identify food components which can be incorporated into the diet to provide evidence-based health benefits, enhancing the functional

Foods, Nutrients and Food Ingredients with Authorised EU Health Claims. http://dx.doi.org/10.1016/B978-0-08-100922-2.00013-9
Copyright © 2018 Elsevier Ltd. All rights reserved.

properties of food products. Nutraceuticals such as arabinoxylans, derived from natural food sources, are likely to deliver health benefits and are also associated with fewer side effects than many pharmacological agents.

13.2 Characterisation of arabinoxylan

13.2.1 Chemical structure and prevalence in major cereal grains

Arabinoxylans are hemicellulose NSPs found in the bran and aleurone (the layer surrounding the endosperm) of many cereal grains. They have a β1→4 linked xylose backbone (1500–5000 units in length) bearing arabinose side chains (Hobden et al., 2015) (Fig. 13.1). Acetic acid, hydroxycinnamic acid, ferulic acid and *p*-coumaric acid are also covalently linked to some of the xylose residues, providing arabinoxylans with antioxidant activity. As a group, arabinoxylans can be classified as either water soluble or water insoluble. Water binding depends on the position and extent of their arabinose substitution; in the absence of arabinose, arabinoxylans bind less water and become much less soluble. The enzymatic hydrolysis of arabinoxylans into shorter chain lengths (up to 60 units) generates arabinoxylo-oligosaccharides (AXOS) and xylo-oligosaccharides (XOS).

Arabinoxylans are found in most cereal grains including wheat, rye, barley, oat, rice and sorghum. Significant levels of arabinoxylans can also be found in psyllium husk, pangola grass, bamboo shoots, corn hulls and banana peel (Muralikrishna and Rao, 2007).

Arabinoxylans constitute 60%–69% of the NSP in wheat bran (where they are primarily in the form of insoluble fibre) and 88% of the NSP in wheat endosperm (primarily in the form of soluble fibre) (Knudsen et al., 1993). After its extraction from

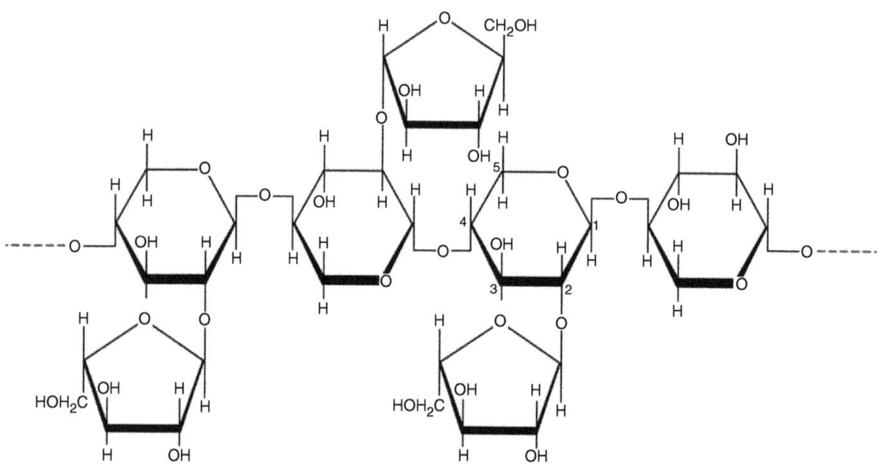

Figure 13.1 Arabinoxylan consists of a xylose backbone with arabinose attached randomly by 1α→2 and/or 1α→3 linkages to the xylose units throughout the chain. *Source*: Diagram by Brendan D. Kellow.

the cereal grain, whole grain wheat flour retains 5%–8% arabinoxylan, whereas highly refined wheat flour contains only 2%–3% arabinoxylan (Hemalatha et al., 2012).

13.2.2 Functional properties of arabinoxylans

Arabinoxylans have many functional properties that are important for bread making and which contribute to their utility as neutraceuticals incorporated in functional foods. Arabinoxylans enhance the bread-making quality of flour by improving the water absorption capacity of dough; increasing dough stability time, mixing tolerance and viscosity; enhancing crumb texture and prolonging shelf life by reducing bread staling and starch retrogradation (Saeed et al., 2011). Water-extractable arabinoxylans additionally increase the elasticity of bread dough, allowing for better gas retention which in turn improves crumb firmness, structure and loaf volume.

13.3 Proposed health benefits from the consumption of arabinoxylans

13.3.1 Digestion of arabinoxylan in the large intestine

As the human intestinal tract does not produce the enzymes necessary to digest arabinoxylans, almost 100% of arabinoxylans consumed reach the colon intact where they acts as a fermentation substrate for colonic bacteria. In vitro and in vivo studies indicate that the presence of arabinoxylans stimulates the colonic growth of specific bacterial genera including the *Bacteroides*, *Roseburia* and *Lactobacillus* (Chassard et al., 2007; Maki et al., 2012; Walton et al., 2012). These bacteria contain arabinoxylan-degrading enzymes such as the endo-β-1,4-xylanases which are able to hydrolyse arabinoxylans into AXOS. These AXOS end products, unlike the intact arabinoxylans, are able to support robust growth of bifidobacteria within the colon, which is postulated to have various health benefits (Grootaert et al., 2009) such as supporting bacterial growth. The fermentation of intact arabinoxylan in the colon also generates short-chain fatty acids (SCFAs), particularly butyrate, acetate and propionate (Damen et al., 2012; Van den Abbeele et al., 2013). The postulated health benefits of SCFAs are discussed in Section 13.3.2.

Most of the arabinoxylan consumed undergoes colonic fermentation. Animal studies indicate that while 85%–100% of ingested arabinoxylan reaches the terminal ileum, only 15% is then recovered in the faeces (Belobrajdic et al., 2012; Glitso et al., 2000). The health benefit of the excreted non-fermentable arabinoxylan remains uncertain, although like other insoluble dietary fibres, it may increase stool water content and reduce intestinal transit time (Walton et al., 2012).

13.3.2 Beneficial effects from SCFAs generated by arabinoxylan fermentation

Production of SCFAs by gut microbial fermentation of arabinoxylan lowers the colonic pH. This changes the colonic environment to favour the proliferation of carbohydrate-degrading bacteria. At the same time, growth of protein-fermenting bacteria is reduced, limiting the accumulation of toxic metabolites, particularly ammonia. The SCFA acetate increases colonic blood flow, while production of butyrate is critical as the principal energetic fuel used by colonocytes (Scheppach, 1994). Butyrate also has important effects promoting epithelial barrier function. Firstly, it enhances the expression of nuclear receptor PPARgamma in colonic epithelial cells which in turn has many flow-on effects: attenuating production of cytokines and other inflammatory factors and maintaining health of epithelial cells lining the colon to reduce gastrointestinal permeability (Eun et al., 2006). Secondly, butyrate upregulates the mucin-associated genes responsible for generating production of the thick mucin layer that plays an important role in maintaining the integrity of the intestinal mucosal barrier (Canani et al., 2011). Arabinoxylan derived bacterial fermentation products also upregulate glucagon-like peptide 2 (GLP-2) production, which supports gut health by promoting the proliferation of intestinal crypt cells (Cani et al., 2012).

13.3.3 Effects on appetite regulation

In solution, arabinoxylan acts as a highly viscous fibre, which can delay gastric emptying, assisting in the control of appetite and energy intake (Torsdottir et al., 1991). The SCFAs generated by bacterial fermentation of arabinoxylans can act as ligands for free fatty acid receptor-2 (FFAR-2) and FFAR-3, which are found in tissues such as gut epithelium, adipose tissue, immune cells and skeletal muscle. Binding of SCFAs to these FFARs stimulates the release of many hormones involved in appetite regulation such as peptide YY, glucagon-like peptide-1 (GLP-1) and leptin (Chambers et al., 2015). These hormones can also influence glucose metabolism.

13.3.4 Effects on glucose metabolism

The high viscosity of arabinoxylan slows post-prandial glucose absorption and subsequent insulin secretion by stimulating the formation of an unstirred water layer above the mucin layer lining of the digestive tract. The presence of this unstirred layer slows diffusion of substrates towards the intestinal wall, thereby reducing the rate of glucose transport to the enterocytes (Lafiandra et al., 2014). Additionally, SCFAs generated by bacterial fermentation of arabinoxylan stimulate the production of the intestinal hormones Peptide YY and GLP-1, which slow gastric emptying, reducing the rate at which the carbohydrate substrates that generate glucose reach the intestine. GLP-1 also potentiates glucose-stimulated insulin secretion, which in turn promotes glycogen synthesis (Delzenne et al., 2007), attenuating post-prandial glycaemia in another way. The SCFA propionate also has metabolic benefits by inhibiting lipogenesis in the liver (Wolever et al., 1995).

13.3.5 Effects on lipid metabolism

Arabinoxylan feeding trials have reported some impacts on circulating cholesterol and triglyceride concentrations, although further research is needed in this area to substantiate findings. One study found that the SCFA propionate may inhibit cholesterol synthesis by reducing the activity of 3-hydroxy-3-methylglutaryl-CoA (HMG-CoA) reductase (Roberfroid et al., 2010). Bacterial fermentation of arabinoxylan might also attenuate cholesterol and triglyceride production by stimulating the synthesis of *cis*-9, *trans*-11 conjugated linoleic acid (CLA). While this isoform of CLA reduces cholesterol and triglyceride concentrations in animal studies (Baddini Feitoza et al., 2009), the results of human trials are less conclusive. Finally, arabinoxylan fermentation supports bacterial action in the deconjugation of bile acids. This prevents their enterohepatic recirculation, forcing the liver to remove more cholesterol from the circulation for bile acid production (Fava et al., 2006).

13.3.6 Enhanced mineral absorption

The presence of SCFAs generated by the microbial fermentation of arabinoxylan increases colonic acidity. This, in turn, enhances the solubility of luminal calcium and magnesium and improves their absorption (Lutz and Scharrer, 1991). Supplementation of the diet with arabinoxylan derived from corn was found to enhance the absorption of calcium and magnesium in the caecum of rats (Lopez et al., 1999).

13.3.7 Antioxidant and anti-cancer activity

Ferulic acid is a natural polyphenol capable of neutralising reactive oxygen species and inducing the destruction of cancerous cells. The ferulic acid component of arabinoxylans has been found to exhibit antioxidant and anti-tumour properties in vivo (Ferguson et al., 2005), but research trials are required to determine whether these properties also occur in humans.

In vitro studies suggest that a rice arabinoxylan supplement may be capable of acting as a biological response modifier, sensitising cancer cells to apoptosis and chemotherapeutic agents (Ghoneum and Gollapudi, 2003; Gollapudi and Ghoneum, 2008). Animal and in vitro studies also indicate that arabinoxylans may exert immunomodulatory effects, including stimulation of natural killer cell activity (Ogawa et al., 2005; Perez-Martinez et al., 2015).

13.4 Authorised EU health claim

13.4.1 Legislation in Europe

The European Food Safety Authority (EFSA) has concluded there is sufficient scientific evidence to substantiate a health claim that 'Consumption of arabinoxylan contributes to a reduction of the glucose rise after a meal' (EFSA, 2011). The Panel recognised an established cause-and-effect relationship between the consumption of arabinoxylan from wheat endosperm and the reduction of post-prandial glycaemic responses (as long as post-prandial insulinaemic responses are not disproportionally increased), and concluded that this may have a beneficial physiological effect.

This conclusion has been supported by other authorities. Further to publication of the positive opinion from EFSA (2011), the European Union authorised use of the health claim with the following wording: 'Consumption of arabinoxylan as part of a meal contributes to a reduction of the blood glucose rise after that meal'.

13.4.2 Conditions of use and target population

The conditions of use for the claim state that 8 g of arabinoxylan-rich fibre produced from wheat endosperm (comprising at least 60% arabinoxylan by weight) per 100 g of available carbohydrates in a quantified portion as part of the meal should be consumed to obtain the claimed effect. Consumers should be informed that the beneficial effect is obtained by consuming the arabinoxylan-rich fibre from wheat endosperm as part of a meal.

The target population are individuals who wish to reduce their post-prandial glycaemic responses. It is recognised that reduction of post-prandial glucose excursions will be most beneficial to consumers with disturbed metabolic responses, that is consumers with impaired glucose tolerance, pre-diabetes or type 2 diabetes. Here the control of post-prandial glycaemic excursions can reduce the risk of diabetic complications including the risk of heart disease and microvascular degenerative changes, leading to conditions such as retinopathy and nephropathy. However, this claim of long-term health benefit still requires substantiation with long-term prospective studies. Also, EU health claims are not specifically targeted at population groups with diseases.

Consumption of food products containing arabinoxylan can lower the glycaemic index of a meal and may enhance satiety in individuals trying to maintain or lose weight (Ley et al., 2014). As elevated post-prandial blood glucose levels are an independent predictor of increased CVD risk in people with diabetes (Cavalot et al., 2011), consumption of arabinoxylan-rich foods to lower post-prandial glucose and insulin responses appears a useful option for consumers aiming to improve their cardiovascular health. However, the current claim is authorised under Article 13.1 (general function claims) of the EU Health Claims Regulation and addresses only the effect of arabinoxylan on the blood glucose rise after a meal, and not its association with risk of disease. This would require a separate application under Article 14.1(a) for a reduction of disease risk claim (Corrigenda, 2007).

13.4.3 Substantiating evidence from human studies

Eight controlled trials exploring the effect of dietary arabinoxylan supplementation on human glucose and insulin homeostasis have been published to date (Table 13.1). Of these studies, four were carried out in healthy adults (Cloetens et al., 2010; Lu et al., 2000; Lappi et al., 2013; Mohlig et al., 2005), two involved individuals with the metabolic syndrome (Hartvigsen et al., 2014a,b), one was undertaken in people with impaired glucose tolerance (pre-diabetes) (Garcia et al., 2006, 2007) and one in subjects with type 2 diabetes (Lu et al., 2004). All eight studies were based on small sample sizes ($\leq$20 participants) and were of short duration, with treatment periods ranging from one single meal to an intervention of 6-week duration. Methodological

Table 13.1 Summary of published human studies investigating the effect of dietary arabinoxylan intake on blood glucose and insulin levels

Studies	Subjects	AX sources	Treatments	Results	References and quality scores
Healthy subjects					
Acute RCT crossover	15 Subjects (mean age: 57 years; 60% F) Mean BMI: 26 kg/m^2	Rye	Single test meal containing: • WWB (control) • WWB plus rye bran • WWB plus bio-processed rye bran (with 8.3 g total AX, 3.8 g soluble AX) • Wholegrain rye bread Took urine at 0,4,8,12 h Six blood samples over 4 h	WWB with AX: • Increased urinary excretion of FA especially from 0–4 h • *No effect on glucose and insulin responses* • No effect on excretion of benzoic, phenylpropionic and phenylacetic acid metabolites	Lappi et al. (2013) Quality score: 7/14 (moderate)
Acute controlled crossover	15 Subjects (mean age: 26 years; 60% F) Only 11 with NGT (64% F)	Wheat	Given breakfast: • Control white bread • AX-enriched (6 g) white bread Order of intervention not randomised	In the entire cohort: • *AX breakfast did not change 2-h glucose but decreased insulin* In those with NGT: • *AX did not change 2-h glucose and only a trend towards lower insulin* • Increase in 2-h plasma ghrelin	Mohlig et al. (2005) Quality score: 6/14 (moderate)
Acute RCT crossover	14 Subjects (mean age: 32 years; 64% F) Mean BMI: 23 kg/m^2	Wheat	Isoenergic breakfasts (bread, margarine, jam = 75-g available CHO, 10-g protein, 14-g fat) and with: • 0 g AX-rich fiber	Versus breakfast with 0-g AX, the 6- or 12-g AX breakfast gave: • *Lower peak postprandial glucose*	Lu et al. (2000) Quality score: 9 (high)

(*Continued*)

Table 13.1 Summary of published human studies investigating the effect of dietary arabinoxylan intake on blood glucose and insulin levels (*cont.*)

Studies	Subjects	AX sources	Treatments	Results	References and quality scores
			• 6 g total AX-rich fiber, <2.5 g soluble AX • 12 g total AX-rich fiber, <5 g soluble AX 3-day washout	• *Lower IAUC for glucose and insulin* (Testing in 30 volunteers showed bread with AX was as palatable as the bread without AX)	
Placebo-controlled RCT crossover	20 Subjects (mean age: 24 years; 70% F) Mean BMI: 21 kg/m^2	Wheat	Consumed for 3 weeks: • 10 g/day AXOS • Or placebo (maltodextrin) 4-week wash-out period. Before and after each Tx: • Blood • GI questionnaire • 48 h urine (also at 2 weeks, $n = 10$) • 72 h faeces (also at 2 weeks, $n = 10$)	Consumption of AXOS: • No effect on GI symptoms (except for mild increase in flatulence). • Urinary *p*-cresol (indicator of protein fermentation) decreased after 2 weeks • bifidobacteria increased after 2 weeks (versus after 3 weeks on placebo) • *No change in fasting blood glucose or lipids*	Cloetens et al. (2010) Quality score: 9 (high)
Subjects with metabolic disorders					
Acute RCT crossover	15 With MetS (mean age: 64 years; 45% F) Mean BMI: 31 kg/m^2	Wheat	50-g Available CHO in breakfast of: • Semolina porridge (control) • Porridge plus 4.3 g AX	Semolina porridge with AX plus rye kernels: • *Reduced acute glucose and insulin responses* versus other breakfasts	Hartvigsen et al. (2014a) Quality score: 8/14 (high)

			• Porridge plus rye kernels • Porridge plus 4.6 g AX plus rye kernels • Standard lunch after 4 h • Blood taken over 6 h • Appetite score and breath H_2 every 30 min	Porridge with AX: • *Reduced acute glucose response* • Both AX and AX with rye kernels increased serum butyrate and acetate at 6 h • *There were no differences in responses to a second meal for glucose, insulin,* FFA, GLP-1 or ghrelin	
Acute RCT crossover	15 with MetS (mean age: 64 years; 45% F) Mean BMI: 31 kg/m^2	Wheat	50-g Available CHO in: • Wheat bread (control) • Wheat bread with 7.1-g concentrated AX • Wheat bread with β-glucan • Rye bread with kernels Blood samples over 270 min Appetite score every 30 min Ad libitum E intake at 270 min	Versus wheat bread alone: • *β-Glucan and rye but not AX caused lower initial glycaemic responses* • *AX reduced the glucose peak value* • *Rye-reduced insulin and GIP responses more than AX or* • *β-Glucan* • *β-Glucan lowered insulin responses more than AX* • All test breads increased satiety but all had no effect on subsequent E intake	Hartvigsen et al. (2014b) Quality score: 8/14 (high)
Single-blind RCT crossover	11 Subjects with IGT (mean age: 56 years; 64% F) Mean BMI: 30 kg/m^2	Wheat	Consumed for 6 weeks: • White bread rolls • White bread rolls with 15-g AX • 6-Week washout period	AX Tx resulted in: • *Lower fasting blood glucose*, TAG and apolipoprotein A-1 • *No effect on fasting insulin*, adiponectin, leptin, resistin, apolipoprotein B or FFA	Garcia et al. (2006) Quality score: 9/14 (high)

(*Continued*)

Table 13.1 Summary of published human studies investigating the effect of dietary arabinoxylan intake on blood glucose and insulin levels (*cont.*)

Studies	Subjects	AX sources	Treatments	Results	References and quality scores
Single-blind RCT crossover	11 Subjects with IGT (mean age: 56 years; 64% F) Mean BMI: 30 kg/m^2	Wheat	Consumed for 6 weeks: • White bread rolls • White bread rolls with 15-g AX • 6-Week washout period. Before and after Tx given an LMCT with blood taken over 240 min	Post LMCT the AX Tx gave: • *Lower glucose, insulin* and TAG • Reduced total plasma ghrelin • No difference in plasma acylated ghrelin	Garcia et al. (2007) Quality score: 9/14 (high)
Single-blind RCT crossover	15 subjects with T2D, (Mean age: 60 years, 60% F) mean BMI: 28 kg/m^2 mean duration of T2DM: 2.3 years	Wheat	For 5-weeks ate usual diet plus: • Bread and muffins (50% whole wheat and 50% white flour) • AX-bread and muffins (50% whole wheat, 36% white flour, 14% AX fibre)(adding mean 15.1(95% CI:12.0–18.5) g/day DF) • After Tx-12 subjects gave 24 h faecal sample	AX diet: • Increased faecal output by 61.5 (95%CI:0.2-122.8) g/day [wet weight] • *Lowered FBG, 2 h plasma glucose, 2 h insulin* and serum fructosamine • No effect on blood lipids, body weight, fat mass or blood pressure	Lu et al. (2004) Quality score: 9/14 (high)

AX, Arabinoxylan; AXOS, arabino-oligosaccharides; BG, blood glucose; CHO, carbohydrate; CI, confidence interval; DF, dietary fibre; E, energy; FA, fatty acids; FBG, fasting blood glucose; FFA, free fatty acids; GI, gastrointestinal; GIP, gastric inhibitory peptide; GLP-1, glucagon-like peptide; IAUC, incremental area under the curve; IGT, impaired glucose tolerance; LMCT, liquid meal challenge test; MetS, metabolic syndrome; NGT, normal glucose tolerance; RCT, randomised controlled trial; SCFA, short-chain fatty acids; TAG, triacylglycerol; T2D, type 2 diabetes; Tx, treatment; WWB, white wheat flour bread quality of these trials was variable. Limitations included failure to report the degree of participant and researcher blinding, absence of intention-to-treat analyses, failure to randomise participants adequately to intervention and control treatments, absence of sample size calculations or failure to justify sample size, inclusion of individuals with highly variable glucose tolerance, results with wide confidence intervals and failure to report whether intervention and control meals/diets were matched for total energy content and macronutrients. Some studies also failed to indicate what proportion of soluble versus insoluble arabinoxylan fibre was used in their intervention. This makes it difficult to draw conclusions from the data regarding the hypoglycaemic efficacy of arabinoxylan, as only the soluble fraction exhibits glucose-lowering properties. Single-meal intervention studies are likely to be of limited relevance when considering the physiological effects of long-term arabinoxylan consumption, particularly as Hartvigsen et al. (2014a,b) have found that the initial glucose-lowering response of a single meal containing arabinoxylan failed to be repeated after a second subsequent meal. Two studies (Cloetens et al., 2010; Lappi et al., 2013) found no effect of arabinoxylan fibre consumption on fasting blood glucose in comparison to controls, but these findings are not relevant to the current EFSA health claim, which is only concerned with effects of arabinoxylan fibre on post-prandial glucose excursions.

Assessment of each trial using the Heyland Methodological Quality Criteria (Heyland et al., 2001) indicated that evidence quality ranged from moderate to high. The Heyland Methodological Quality Score for individual studies is based on nine quality criteria: use of randomisation, blinding, intention-to-treat analysis, selection of consecutive eligible patients, comparable groups at baseline, extent of patient follow-up, well-described treatment protocol, equal application of co-interventions across groups, and objectively defined outcomes. The maximum possible score is 14, with studies scoring ≥8 considered to be of high methodological quality (Table 13.1).

Single-meal arabinoxylan fibre-feeding studies in healthy individuals have produced contradictory findings. Two trials (Lappi et al., 2013; Mohlig et al., 2005) of moderate methodological quality reported no effect of arabinoxylan consumption on post-prandial glucose and insulin responses in comparison with control meals, whereas one high methodological quality trial (Lu et al., 2000) found significantly reduced glucose and insulin responses after consumption of arabinoxylan-fortified bread. Based on the limited short-term data currently available in healthy adults, the effect of arabinoxylan fibre on glucose and insulin responses seems to be less pronounced in people with normal glucose tolerance, and is likely to be of limited value for the purpose of glucose control in metabolically healthy individuals.

Two high-quality single-meal studies performed in the same group of 15 individuals with the metabolic syndrome found variable glucose and insulin responses to consumption of porridge (Hartvigsen et al., 2014a) or bread (Hartvigsen et al., 2014b) supplemented with either arabinoxylan, rye kernels, β-glucan or a combination of arabinoxylan plus rye kernels. Foods supplemented with these fibre sources elicited reduced glucose and insulin responses in comparison with standard control porridge/ bread, but major differences in the macronutrient composition of each treatment make the findings difficult to interpret. For example the arabinoxylan-supplemented bread contained twice the amount of protein as each of the other breads tested, potentially altering the rate of starch digestion.

Two high-quality feeding trials have been performed in individuals with impaired glucose tolerance (Garcia et al., 2007) and type 2 diabetes (Lu et al., 2004). In the first study, 11 overweight subjects with impaired glucose tolerance participated in a single-blind randomised controlled crossover trial, receiving either 15 g of arabinoxylan (approximately 8 g soluble fibre) per day or placebo for 6 weeks each. Intervention periods were separated by a 6-week washout. Following each intervention, participants underwent a liquid meal challenge test containing 74 g of carbohydrate. In comparison with placebo, arabinoxylan consumption resulted in significant reductions in serum glucose and insulin responses. The study was appropriately powered to detect these post-prandial changes (Garcia et al., 2007). The second study involved 15 overweight subjects with type 2 diabetes, who participated in a single-blind randomised controlled crossover study, receiving either 15 g of arabinoxylan (approximately 9 g soluble fibre) per day or placebo for 5 weeks each. There was no washout between interventions and post-prandial glucose and insulin were measured following the administration of a 75 g oral glucose tolerance test. The 2-h glucose and insulin concentrations after the oral glucose tolerance test were significantly lower after the arabinoxylan diet than after the control diet, and the study was sufficiently powered to detect these changes (Lu et al., 2004).

For the purposes of assessing the scientific evidence to substantiate the arabinoxylan health claim, the EFSA Panel on Dietetic Products, Nutrition and Allergies considered the studies performed by Lu et al. (2000, 2004): one single-meal study carried out in 14 healthy individuals, and a 5-week study involving 15 subjects with type 2 diabetes. The Panel concluded that the single-meal study in healthy subjects showed a dose–response relationship between the intake of arabinoxylan produced from wheat endosperm and reduction in post-prandial glycaemic and insulinaemic responses. While the Panel noted that the findings obtained from the 5-week study in people with type 2 diabetes did not result from the direct measurement of post-prandial glycaemic responses to arabinoxylan consumption (but were performed after an oral glucose load), they concluded that a cause-and-effect relationship has been established between the consumption of arabinoxylan produced from wheat endosperm and reduction of post-prandial glycaemic responses. The Panel noted that the mechanism by which soluble fibres (including arabinoxylan produced from wheat endosperm) could exert the claimed effect is well established, and relates to the increased viscosity of the meal bolus, which delays the rate of absorption of nutrients (including glucose) in the small intestine (EFSA, 2011). Although not specifically noted by the EFSA Panel, AXOS resulting from the hydrolysis of arabinoxylan has demonstrated prebiotic properties (Broekaert et al., 2011), which is another mechanism whereby arabinoxylan may attenuate post-prandial glucose excursions. Prebiotics are defined as non-digestible food ingredients that beneficially affect the host by selectively stimulating the growth and/or activity of one or a limited number of bacteria in the colon, thereby improving host health (Gibson and Roberfroid, 1995). Ingestion of prebiotics typically shifts the composition of the intestinal microbiota towards a relative increase in beneficial bacterial species, resulting in the increase in SCFA production in the colon, which has been shown in a meta-analysis to reduce post-prandial glucose and insulin responses (Kellow et al., 2014).

13.5 Conditions of use for arabinoxylan and safety issues

13.5.1 Recommended dose

It is estimated from a western diet that the average daily consumption of arabinoxylan in the absence of fibre supplements ranges from 1–5 g/day. To obtain a beneficial effect on post-prandial glucose excursions, the conditions of use for the claim are that 8 g of arabinoxylan-rich fibre derived from wheat endosperm (which is at least 60% arabinoxylan by weight) per 100 g of available carbohydrates should be consumed per quantified portion as part of a meal.

13.5.2 Safety issues

A 90-day rat feeding study has indicated that consumption of 4.4 g of wheat bran extract (rich in arabinoxylan) per kilogram body weight per day was not associated with any adverse effects. In addition, in vitro genotoxicity assays on wheat bran extract showed no evidence of mutagenic or clastogenic activity (Francois et al., 2010). Wheat bran extract is Generally Recognised As Safe (GRAS) by the US Food and Drug Administration (FDA).

Consumption of 30 g/day of wheat bran extract (comprising ≤70% AXOS) did not increase symptoms of gastrointestinal distress in healthy individuals (Francois et al., 2014). However, a mild increase in flatulence has been reported in people consuming 10–15 g of AXOS/day (Cloetens et al., 2010). This effect can usually be minimised by gradually increasing the daily dose of dietary fibre over a 1–2 week period so that the body can adapt.

13.5.3 Upper level of intake

To date no upper level of intake for arabinoxylan has been set. Potentially very large quantities of arabinoxylan may displace other nutrients from the diet and/or affect absorption of important minerals such as iron, calcium and zinc. Studies on the human consumption of wheat bran conclude that adverse effects on mineral absorption are unlikely within amounts customarily eaten (16–20 g/day) (Sandberg et al., 1982). However, phytic acid, which is present in all cereal grains, may suppress mineral bioavailability through the formation of insoluble metal phytate complexes.

Some arabinoxylan supplements (e.g. MGN-3, BioBran) may contain high levels of calcium and phosphorus, which may be harmful for individuals with impaired renal function. People with diabetes who are taking blood glucose-lowering medications and/or insulin should be made aware of the hypoglycaemic effects of arabinoxylan consumption, and that some reduction in their medication dose may be required to avoid hypoglycaemia.

Table 13.2 Estimated soluble arabinoxylan content of common foods

Food items	Portion sizes	Estimated soluble arabinoxylan contents (g)
Kelloggs all bran cereal	0.5 cup	0.7
Quick oats	1 cup	1.5
Oat bran	0.5 cup	1.6
Rice bran	0.5 cup	1.2
Unprocessed wheat bran	0.5 cup	0.9
Wholegrain bread	1 slice	1.0
Rye bread	1 slice	0.6
White bread	1 slice	0.2
White pasta, cooked	1 cup	0.9
Brown rice, cooked	1 cup	1.2
Psyllium husk	10 g	5.7
Metamucil powder	10 g	5.0
Naxus arabinoxylan fibre	10 g	5.0

13.6 Consumer issues

13.6.1 Arabinoxylan content of common foods

Arabinoxylans are naturally present in most grain-based food and beverages including breads, pastry, pasta, biscuits, crackers, pancakes, waffles, ready-to-eat breakfast cereals, cereal bars and beer (Table 13.2). Arabinoxylan is palatable and arabinoxylan-rich foods are well accepted by consumers (Garcia et al., 2007). Arabinoxylan supplements have been successfully added to breads, porridge, chapati and biscuits in clinical trials.

In the absence of dietary fibre supplements, it will be difficult for consumers to obtain 8 g of soluble arabinoxylan fibre per meal from natural food sources. Individuals following reduced kilojoule diets; those with poor appetites and with moderate–low carbohydrate intakes will be particularly disadvantaged. The introduction of arabinoxylan fibre supplements and food products fortified with arabinoxylan fibre will be beneficial for these consumers.

13.7 Conclusions

The consumption of arabinoxylan fibre for reducing post-prandial blood glucose excursions in people with dysfunctional glucose metabolism is supported by high-quality, short-term clinical trials in humans. To achieve the recommended intake of 8 g of arabinoxylan-rich fibre from wheat endosperm, consumption of fortified foods or arabinoxylan fibre supplements will be required.

13.8 Sources of further information and advice

Corrigenda (2007) Regulation (EC) No. 1924/2006 of the European Parliament and of the Council of 20 December 2006 on nutrition and health claims made on foods. Off. J. Eur. Union, L12, 3–18.

References

Aldoori, W., Giovannucci, E., Rockett, H., Sampson, L., Rimm, E., Willett, W., 1998. A prospective study of dietary fiber types and symptomatic diverticular disease in men. J. Nutr. 128 (4), 714–719.

Baddini Feitoza, A., Fernandes Pereira, A., Ferreira da Costa, N., Goncalves Ribeiro, B., 2009. Conjugated linoleic acid (CLA): effect modulation of body composition and lipid profile. Nutr. Hosp. 24 (4), 422–428.

Beagley, J., Guariguata, L., Weil, C., Motala, A., 2014. Global estimates of undiagnosed diabetes in adults. Diabetes Res. Clin. Pract. 103 (2), 150–160.

Belobrajdic, D., Bird, A., Conlon, M., Williams, B., Kang, S., McSweeney, C., Zhang, D., Bryden, W., Gidley, M., Topping, D., 2012. An arabinoxylan-rich fraction from wheat enhances caecal fermentation and protects colonocyte DNA against diet-induced damage in pigs. Br. J. Nutr. 107 (9), 1274–1282.

Broekaert, W., Courtin, C., Verbeke, K., Van de Wiele, T., Verstraete, W., Delcour, J., 2011. Prebiotic and other health-related effects of cereal-derived arabinoxylans, arabinoxylan-oligosaccharides, and xylooligosaccharides. Crit. Rev. Food Sci. Nutr. 51 (2), 178–194.

Canani, R., Costanzo, M., Leone, L., Pedata, M., Meli, R., Calignano, A, 2011. Potential beneficial effects of butyrate in intestinal and extraintestinal diseases. World J. Gastroenterol. 17 (12), 1519–1528.

Cani, P., Osto, M., Geurts, L., Everard, A., 2012. Involvement of gut microbiota in the development of low-grade inflammation and type 2 diabetes associated with obesity. Gut Microbes 3 (4), 279–288.

Cavalot, F., Pagliarino, A., Valle, M., Di Martino, L., Bonomo, K., Massucco, P., Anfossi, G., Trovati, M., 2011. Postprandial blood glucose predicts cardiovascular events and all-cause mortality in type 2 diabetes in a 14-year follow-up: lessons from the San Luigi Gonzaga Diabetes Study. Diabetes Care 34 (10), 2237–2243.

Chambers, E., Morrison, D., Frost, G, 2015. Control of appetite and energy intake by SCFA: what are the potential underlying mechanisms? Proc. Nutr. Soc. 74 (3), 328–336.

Chassard, C., Goumy, V., Leclerc, M., Del'homme, C., Bernalier-Donadille, A., 2007. Characterization of the xylan-degrading microbial community from human faeces. FEMS Microbiol. Ecol. 61 (1), 121–131.

Cloetens, L., Broekaert, W., Delaedt, Y., Ollevier, F., Courtin, C., Delcour, J., Rutgeerts, P., Verbeke, K., 2010. Tolerance of arabinoxylan-oligosaccharides and their prebiotic activity in healthy subjects: a randomised, placebo-controlled cross-over study. Br. J. Nutr. 103 (5), 703–713.

Corrigenda, 2007. Regulation (EC) No. 1924/2006 of the European Parliament and of the Council of 20 December 2006 on nutrition and health claims made on foods. Off. J. Eur. Union, L12, 3–18.

Damen, B., Cloetens, L., Broekaert, W., Francois, I., Lescroart, O., Trogh, I., Arnaut, F., Welling, G., Wijffels, J., Delcour, J., et al., 2012. Consumption of breads containing in situ-produced arabinoxylan oligosaccharides alters gastrointestinal effects in healthy volunteers. J. Nutr. 142 (3), 470–477.

Delzenne, N., Cani, P., Neyrinck, A., 2007. Modulation of glucagon-like peptide 1 and energy metabolism by inulin and oligofructose: experimental data. J. Nutr. 137 (11 Suppl.), 2547S–2551S.

EFSA, 2011. Scientific Opinion on the substantiation of health claims related to arabinoxylan produced from wheat endosperm and reduction of post-prandial glycaemic responses (ID 830) pursuant to Article 13(1) of Regulation (EC) No. 1924/2006. EFSA J. 9 (6), 2205.

Eun, C., Han, D., Lee, S., Paik, C., Chung, Y., Lee, J., Hahm, J., 2006. Attenuation of colonic inflammation by PPARgamma in intestinal epithelial cells: effect on Toll-like receptor pathway. Dig. Dis. Sci. 51 (4), 693–697.

Fava, F., Lovegrove, J., Gitau, R., Jackson, K., Tuohy, K., 2006. The gut microbiota and lipid metabolism: implications for human health and coronary heart disease. Curr. Med. Chem. 13 (25), 3005–3021.

Ferguson, L., Zhu, S., Harris, P., 2005. Antioxidant and antigenotoxic effects of plant cell wall hydroxycinnamic acids in cultured HT-29 cells. Mol. Nutr. Food Res. 49 (6), 585–593.

Francois, I., Lescroart, O., Veraverbeke, W., Kubaszky, R., Hargitai, J., Esdaile, D., Beres, E., Soni, M., Cockburn, A., Broekaert, W., 2010. Safety assessment of a wheat bran extract containing arabinoxylan-oligosaccharides: mutagenicity, clastogenicity, and 90-day rat-feeding studies. Int. J. Toxicol. 29 (5), 479–495.

Francois, I., Lescroart, O., Veraverbeke, W., Windey, K., Verbeke, K., Broekaert, W., 2014. Tolerance and the effect of high doses of wheat bran extract, containingarabinoxylan-oligosaccharides, and oligofructose on faecal output: a double-blind, randomised, placebo-controlled, cross-over trial. J. Nutr. Sci. 3, e49.

Fuchs, C., Giovannucci, E., Colditz, G., Hunter, D., Stampfer, M., Rosner, B., Speizer, F., Willett, W., 1999. Dietary fiber and the risk of colorectal cancer and adenoma in women. N. Engl. J. Med. 340 (3), 169–176.

Fung, T., Hu, F., Pereira, M., Liu, S., Stampfer, M., Colditz, G., Willett, W., 2002. Whole-grain intake and the risk of type 2 diabetes: a prospective study in men. Am. J. Clin. Nutr. 76 (3), 535–540.

Garcia, A., Otto, B., Reich, S., Weickert, M., Steiniger, J., Machowetz, A., Rudovich, N., Mohlig, M., Katz, N., Speth, M., et al., 2007. Arabinoxylan consumption decreases post-prandial serum glucose, serum insulin and plasma total ghrelin response in subjects with impaired glucose tolerance. Eur. J. Clin. Nutr. 61 (3), 334–341.

Garcia, A., Steiniger, J., Reich, S., Weickert, M., Harsch, I., Machowetz, A., Mohlig, M., Spranger, J., Rudovich, N., Meuser, F., et al., 2006. Arabinoxylan fibre consumption improved glucose metabolism, but did not affect serum adipokines in subjects with impaired glucose tolerance. Horm. Metab. Res. 38 (11), 761–766.

Ghoneum, M., Gollapudi, S., 2003. Modified arabinoxylan rice bran (MGN-3/Biobran) sensitizes human T cell leukemia cells to death receptor (CD95)-induced apoptosis. Cancer Lett. 201 (1), 41–49.

Gibson, G., Roberfroid, M., 1995. Dietary modulation of the human colonic microbiota: introducing the concept of prebiotics. J. Nutr. 125 (6), 1401–1412.

Glitso, L., Mazur, W., Adlercreutz, H., Wahala, K., Makela, T., Sandstrom, B., Bach Knudsen, K., 2000. Intestinal metabolism of rye lignans in pigs. Br. J. Nutr. 84 (4), 429–437.

Gollapudi, S., Ghoneum, M., 2008. MGN-3/Biobran, modified arabinoxylan from rice bran, sensitizes human breast cancer cells to chemotherapeutic agent, daunorubicin. Cancer Detect. Prev. 32 (1), 1–6.

Grootaert, C., Van den Abbeele, P., Marzorati, M., Broekaert, W., Courtin, C., Delcour, J., Verstraete, W., Van de Wiele, T., 2009. Comparison of prebiotic effects of arabinoxylan oligosaccharides and inulin in a simulator of the human intestinal microbial ecosystem. FEMS Microbiol. Ecol. 69 (2), 231–242.

Guariguata, L., Whiting, D., Hambleton, I., Beagley, J., Linnenkamp, U., Shaw, J., 2014. Global estimates of diabetes prevalence for 2013 and projections for 2035. Diabetes Res. Clin. Pract. 103 (2), 137–149.

Hartvigsen, M., Gregersen, S., Laerke, H., Holst, J., Bach Knudsen, K., Hermansen, K., 2014a. Effects of concentrated arabinoxylan and beta-glucan compared with refined wheat and whole grain rye on glucose and appetite in subjects with the metabolic syndrome: a randomized study. Eur. J. Clin. Nutr. 68 (1), 84–90.

Hartvigsen, M., Laerke, H., Overgaard, A., Holst, J., Bach Knudsen, K., Hermansen, K, 2014b. Postprandial effects of test meals including concentrated arabinoxylan and whole grain rye in subjects with the metabolic syndrome: a randomised study. Eur. J. Clin. Nutr. 68 (5), 567–574.

Hemalatha, M., Bhagwat, S., Salimath, P., Prasada Rao, U., 2012. Enhancement of soluble dietary fibre, polyphenols and antioxidant properties of chapatis prepared from whole wheat flour dough treated with amylases and xylanase. J. Sci. Food Agric. 92 (4), 764–771.

Heyland, D., Novak, F., Drover, J., Jain, M., Su, X., Suchner, U., 2001. Should immunonutrition become routine in critically ill patients? A systematic review of the evidence. JAMA 286 (8), 944–953.

Hobden, M., Guerin-Deremaux, L., Rowland, I., Gibson, G., Kennedy, O., 2015. Potential anti-obesogenic properties of non-digestible carbohydrates: specific focus on resistant dextrin. Proc. Nutr. Soc. 74 (3), 258–267.

Jones, J., 2013. Dietary fiber future directions: integrating new definitions and findings to inform nutrition research and communication. Adv. Nutr. 4 (1), 8–15.

Kellow, N., Coughlan, M., Reid, C., 2014. Metabolic benefits of dietary prebiotics in human subjects: a systematic review of randomised controlled trials. Br. J. Nutr. 111 (7), 1147–1161.

Knudsen, K., Jensen, B., Hansen, I., 1993. Digestion of polysaccharides and other major components in the small and large intestine of pigs fed on diets consisting of oat fractions rich in beta-D-glucan. Br. J. Nutr. 70 (2), 537–556.

Lafiandra, D., Riccardi, G., Shewry, P., 2014. Improving cereal grain carbohydrates for diet and health. J. Cereal Sci. 59 (3), 312–326.

Lappi, J., Aura, A., Katina, K., Nordlund, E., Kolehmainen, M., Mykkanen, H., Poutanen, K., 2013. Comparison of postprandial phenolic acid excretions and glucose responses after ingestion of breads with bioprocessed or native rye bran. Food Funct. 4 (6), 972–981.

Ley, S., Hamdy, O., Mohan, V., Hu, F., 2014. Prevention and management of type 2 diabetes: dietary components and nutritional strategies. Lancet 383 (9933), 1999–2007.

Liu, S., Manson, J., Lee, I., Cole, S., Hennekens, C., Willett, W., Buring, J., 2000. Fruit and vegetable intake and risk of cardiovascular disease: the Women's Health Study. Am. J. Clin. Nutr. 72 (4), 922–928.

Lopez, H., Levrat, M., Guy, C., Messager, A., Demigne, C., Remesy, C., 1999. Effects of soluble corn bran arabinoxylans on cecal digestion, lipid metabolism, and mineral balance (Ca, Mg) in rats. J. Nutr. Biochem. 10 (9), 500–509.

Lu, Z., Walker, K., Muir, J., Mascara, T., O'Dea, K., 2000. Arabinoxylan fiber, a byproduct of wheat flour processing, reduces the postprandial glucose response in normoglycemic subjects. Am. J. Clin. Nutr. 71 (5), 1123–1128.

Lu, Z., Walker, K., Muir, J., O'Dea, K., 2004. Arabinoxylan fibre improves metabolic control in people with Type II diabetes. Eur. J. Clin. Nutr. 58 (4), 621–628.

Lutz, T., Scharrer, E., 1991. Effect of short-chain fatty acids on calcium absorption by the rat colon. Exp. Physiol. 76 (4), 615–618.

Maki, K., Gibson, G., Dickmann, R., Kendall, C., Chen, C., Costabile, A., Comelli, E., McKay, D., Almeida, N., Jenkins, D., et al., 2012. Digestive and physiologic effects of a wheat bran extract, arabino-xylan-oligosaccharide, in breakfast cereal. Nutrition 28 (11–12), 1115–1121.

McKeown, N., Meigs, J., Liu, S., Saltzman, E., Wilson, P., Jacques, P., 2004. Carbohydrate nutrition, insulin resistance, and the prevalence of the metabolic syndrome in the Framingham Offspring Cohort. Diabetes Care 27 (2), 538–546.

Mohlig, M., Koebnick, C., Weickert, M., Lueder, W., Otto, B., Steiniger, J., Twilfert, M., Meuser, F., Pfeiffer, A., Zunft, H., 2005. Arabinoxylan-enriched meal increases serum ghrelin levels in healthy humans. Horm. Metab. Res. 37 (5), 303–308.

Muralikrishna, G., Rao, M., 2007. Cereal non-cellulosic polysaccharides: structure and function relationship: an overview. Crit. Rev. Food Sci. Nutr. 47 (6), 599–610.

Ogawa, K., Takeuchi, M., Nakamura, N., 2005. Immunological effects of partially hydrolyzed arabinoxylan from corn husk in mice. Biosci. Biotechnol. Biochem. 69 (1), 19–25.

Pereira, M., O'Reilly, E., Augustsson, K., Fraser, G., Goldbourt, U., Heitmann, B., Hallmans, G., Knekt, P., Liu, S., Pietinen, P., et al., 2004. Dietary fiber and risk of coronary heart disease: a pooled analysis of cohort studies. Arch. Intern. Med. 164 (4), 370–376.

Perez-Martinez, A., Valentin, J., Fernandez, L., Hernandez-Jimenez, E., Lopez-Collazo, E., Zerbes, P., Schworer, E., Nunez, F., Martin, I., Sallis, H., et al., 2015. Arabinoxylan rice bran (MGN-3/Biobran) enhances natural killer cell-mediated cytotoxicity against neuroblastoma in vitro and in vivo. Cytotherapy 17 (5), 601–612.

Reicks, M., Jonnalagadda, S., Albertson, A., Joshi, N., 2014. Total dietary fiber intakes in the US population are related to whole grain consumption: results from the National Health and Nutrition Examination Survey 2009 to 2010. Nutr. Res. 34 (3), 226–234.

Roberfroid, M., Gibson, G., Hoyles, L., McCartney, A., Rastall, R., Rowland, I., Wolvers, D., Watzl, B., Szajewska, H., Stahl, B., et al., 2010. Prebiotic effects: metabolic and health benefits. Br. J. Nutr. 104 (Suppl. 2), S1–S63.

Saeed, F., Pasha, I., Anjum, F., Sultan, M, 2011. Arabinoxylans and arabinogalactans: a comprehensive treatise. Crit. Rev. Food Sci. Nutr. 51 (5), 467–476.

Sandberg, A., Hasselblad, C., Hasselblad, K., Hulten, L, 1982. The effect of wheat bran on the absorption of minerals in the small intestine. Br. J. Nutr. 48 (2), 185–191.

Scheppach, W., 1994. Effects of short chain fatty acids on gut morphology and function. Gut 35 (1 Suppl.), S35–S38.

Torsdottir, I., Alpsten, M., Holm, G., Sandberg, A., Tolli, J., 1991. A small dose of soluble alginate-fiber affects postprandial glycemia and gastric emptying in humans with diabetes. J. Nutr. 121 (6), 795–799.

Van den Abbeele, P., Venema, K., Van de Wiele, T., Verstraete, W., Possemiers, S, 2013. Different human gut models reveal the distinct fermentation patterns of arabinoxylan versus inulin. J. Agric. Food Chem. 61 (41), 9819–9827.

Walton, G., Lu, C., Trogh, I., Arnaut, F., Gibson, G., 2012. A randomised, double-blind, placebo controlled cross-over study to determine the gastrointestinal effects of consumption of arabinoxylan-oligosaccharides enriched bread in healthy volunteers. Nutr. J. 11, 36.

Wolever, T., Spadafora, P., Cunnane, S., Pencharz, P., 1995. Propionate inhibits incorporation of colonic [1,2-13C] acetate into plasma lipids in humans. Am. J. Clin. Nutr. 61 (6), 1241–1247.

Authorised EU health claim for alpha-cyclodextrin

14

Sarah Kuczora
Dudley Metropolitan Borough Council, Dudley, United Kingdom

14.1 Introduction

This chapter provides an overview of the authorised Article 13.1 'general function' health claim for alpha-cyclodextrin and the reduction of post-prandial glycaemic responses which was awarded in June 2013 (EU, 2013). This was one of the claims that fell into the category of 'insufficient evidence' on initial application. Applicants for this group of claims were provided with the opportunity to resubmit the new evidence, and this process is briefly described. The chapter also covers characterisation of alpha-cyclodextrin, a summary of the substantiating evidence, the approved claim wording and conditions of use, and consideration of consumer and product development issues.

The European Food Safety Authority (EFSA) stated that the target population for this claim should be those who want to reduce their post-prandial glycaemic responses. The reduction of post-prandial glucose responses may be beneficial to those with impaired glucose responses which is also referred to as 'pre-diabetes'. Pre-diabetes is common in the general adult population and is associated with an increased risk of chronic diseases such as obesity, type 2 diabetes and cardiovascular disease. However, raised blood glucose responses develop slowly and can remain undetected for some time. Therefore, potential consumers may not be aware of their need for products bearing this health claim. The term 'alpha-cyclodextrin' may also be unfamiliar to consumers which can be a barrier to purchase intentions. Food manufacturers using the claim may wish to draw on previous research considering health claims and consumer perceptions to develop a product and also a marketing strategy that appeals to the target audience. However, in doing so, they must ensure that the conditions of the health claims regulation are still met.

14.2 Process for authorisation of the claim

An initial application, made under the Article 13.1 health claims process, for effect of alpha-cyclodextrin on post-prandial glycaemic responses was deemed by EFSA to provide'insufficient evidence' (EFSA, 2010). This category of opinion was treated distinctly from a positive opinion which established cause and effect, and a negative opinion, where it was deemed that no cause-and-effect relationship had been established (Binns, 2014). As discussed by Sadler (2015), there were a number of

Foods, Nutrients and Food Ingredients with Authorised EU Health Claims. http://dx.doi.org/10.1016/B978-0-08-100922-2.00014-0
Copyright © 2018 Elsevier Ltd. All rights reserved.

claims that fell into this category following the initial round of EFSA assessments and this provided a unique opportunity for a number of applicants to reapply for approval. Briefly, the resubmission was required to contain new evidence that was not previously submitted and for the health claim to be identical to the initial application. In agreement with EFSA, a template was produced to structure the summary of evidence. Applicants received a 4-month period for submission, with a deadline of 30 September 2011. Details of the further assessment process are described in full by Sadler (2015). Additional information for this particular application was submitted by the competent Authority of the Netherlands.

14.3 Characterisation

The food that is the subject of this health claim is alpha-cyclodextrin, which EFSA deemed to be sufficiently characterised. Alpha-cyclodextrin is considered to be a soluble dietary fibre, also known as fermentable fibre. In its opinion EFSA stated that alpha-cyclodextrin (also known as cyclohexaamylose or cyclomaltohexaose) is a cyclic saccharide comprising of six glucose units linked by alpha-1,4 bonds, produced by the action of cyclodextrin glucosyltransferase on hydrolysed starch syrups. Its structure provides a hydrophobic cavity surrounded by a hydrophilic outer surface which makes alpha-cyclodextrin water soluble. Human salivary and pancreatic amylases are unable to hydrolyse alpha-cyclodextrin sufficiently; however, it can be hydrolysed by alpha-amylases of bacterial origin present in the human intestine.

14.4 Target population

The claimed effect is the reduction of post-prandial glycaemic (glucose) responses and the target population are individuals who want to reduce their post-prandial glucose responses. Post-prandial glycaemia is described as the rise in blood glucose concentrations after the consumption of a food or meal. It is a normal physiological response that varies in size and duration and may be influenced by the chemical and physical nature of the food or meal eaten as well as by individual factors (Venn and Green, 2007). EFSA agreed that a reduction in post-prandial blood glucose responses may be beneficial to individuals with impaired glucose tolerance, as long as their post-prandial insulinaemic responses are not disproportionally increased. Impaired glucose tolerance is recognised as being common in the general adult population.

14.5 Authorised claim wording

The claim wording proposed by EFSA in its opinion was 'Consumption of alpha-cyclodextrin contributes to the reduction of the blood glucose rise after starch-containing meals'. However, the final claim wording as it appears in the EU register of claims (http://ec.europa.eu/nuhclaims) is ‘Consumption of alpha-cyclodextrin as part of a starch-containing meal contributes to the reduction of the blood glucose rise after that meal’.

14.6 Substantiating evidence

14.6.1 Initial claim submission

For the initial claim submission (EFSA, 2010), two references were provided, one of which was an unpublished project report that contained the study description but not the results (Diamantis and Bär, 2002). Therefore, no conclusions could be drawn from this study.

The second study (Buckley et al., 2006) was a double-blinded, randomised crossover, human intervention study. In the study, 10 healthy subjects (5 females) consumed boiled white rice, containing 50 g digestible carbohydrates (starch) and either 0 (control), 2, 5 or 10 g alpha-cyclodextrin. Subjects consumed each test meal on a single occasion following an overnight fast, with a 2-day wash-out period. Blood glucose and insulin levels were measured at baseline and over a 2-h period following each meal. The incremental area under the curve (iAUC) was significantly lower for boiled white rice containing 5 g alpha-cyclodextrin (mean ± SEM−20.4 ± 15.4%; $P = 0.03$) or 10 g alpha-cyclodextrin (mean ± SEM−49.6 ± 9.9%; $P = 0.001$) compared with the control rice. No effect on post-prandial insulin levels was observed for either alpha-cyclodextrin dose. EFSA observed that the study demonstrated a dose-dependent effect of alpha-cyclodextrin on the reduction of post-prandial glycaemic responses. However, based on the totality of the evidence presented, EFSA concluded that there was insufficient evidence to establish a cause-and-effect relationship.

14.6.2 Further assessment

Under the framework of further assessment, 11 additional studies were presented in evidence. Of these, seven studies were considered not pertinent to scientific substantiation of the claim. In summary:

- Two human intervention studies (Comerford et al., 2011; Grunberger et al., 2007) considered the effects of alpha-cyclodextrin on health outcomes other than post-prandial glycaemic responses, for example blood lipids, weight loss and body weight
- Two human intervention studies (Ochiai et al., 2008; Shimazu et al., 2009) investigated the effects of an anti-diabetic medication on adiponectin concentrations
- Two papers (Weyer et al., 2001; Yun et al., 2009) detailed the association between plasma adiponectin levels and insulin sensitivity and insulinaemia in obesity and/or type 2 diabetes
- One systematic review (Li et al., 2009) considered the relationship between plasma adiponectin concentrations and type 2 diabetes incidence.

EFSA’s evaluation was based on the four remaining studies provided for the further assessment as well as the study by Buckley et al. (2006) presented for the initial assessment. A summary of the substantiating evidence can be found in Table 14.1.

Table 14.1 Summary of substantiating evidence for alpha-cyclodextrin and the reduction of post-prandial glucose responses

Authors	Study designs	Main aims	Controls	Interventions	Main outcomes
Buckley et al. (2006)	Double-blinded, randomised cross-over, human intervention study	Determine the effect of alpha-cyclodextrin added to white rice on glucose responses	White rice (50-g starch) + 0-g alpha-cyclodextrin	White rice (50-g starch) + 2-, 5- or 10-g alpha-cyclodextrin	↓ Post-prandial glucose responses following 5- or 10-g of alpha-cyclodextrin cf. control
Diamantis and Bär (2002)	Single-blinded, cross-over study, human intervention study	Determine the effect of alpha-cyclodextrin added to white bread on glucose responses	White bread (50-g starch) + 0-g alpha-cyclodextrin	White bread (50- g starch) + 10-g alpha-cyclodextrin or 25- g alpha-cyclodextrin dissolved in 250 mL water	↓ Post-prandial glucose responses following 10-g alpha-cyclodextrin cf. control
Gentilcore et al. (2011)	Double-blinded, randomised cross-over, human intervention study	Determine the effect of alpha-cyclodextrin added to sucrose on glucose responses and gastric emptying time	Sucrose (100 g) in water	Sucrose (100 g) in water + 10-g alpha-cyclodextrin	↔ Post-prandial glucose responses following 10- g alpha-cyclodextrin cf. control ↔ Gastric emptying time following 10-g alpha-cyclodextrin cf. control
Koukiekolo et al. (2001)	In vitro kinetic study	Understand the inhibitory effect of alpha-cyclodextrin on amylase	No control	Alpha-cyclodextrin	Competitive inhibition of amylose hydrolysis demonstrated
Oudjeriouat et al. (2003)	In vitro kinetic study	Determine the inhibitory effect of alpha-cyclodextrin on amylase cf. acarbose (a known strong inhibitor)	Acarbose	Alpha-cyclodextrin	Alpha-cyclodextrin shown to be a weak inhibitor of amylase cf. control

An unpublished report by Diamantis and Bär (2002), detailed the results of the study protocol presented in the initial application. The study aimed to determine the effect of alpha-cyclodextrin on the glycaemic and insulinaemic responses to starch, and was a single-blinded, cross-over study in 12 healthy male subjects (age 23–24 years and mean weight 73.3 kg). After an overnight fast, subjects consumed either 100 g white bread (providing 50 g starch) together with 0 (control) or 10g alpha-cyclodextrin, or 25 g alpha-cyclodextrin dissolved in 250 mL water. The wash-out period was at least 2 days. Baseline blood glucose and serum insulin levels were measured, followed by samples up to 180 min (2.5 h) after the meal. The iAUC for blood glucose concentrations was significantly lower after consumption of white bread together with 10 g alpha-cyclodextrin compared with the control (−58.63%; mean ± SEM 1543.5 ± 469.6 vs. 3731.2 ± 581.9 mg/dL/min; $P < 0.01$). The iAUC was also significantly lower for insulin (−56.48%; mean ± SEM 2559.9 ± 255.2 vs. 5883.5 ± 479.0 μU/mL/min; $P < 0.001$). The glycaemic response of white bread with alpha-cyclodextrin and alpha-cyclodextrin alone was 43 and 4.3%, respectively, relative to that of white bread alone (100%). The insulinaemic response of white bread with alpha-cyclodextrin relative to white bread alone (100%) was 45%. Data was also analysed using a linear mixed model. At T15-T90 for glucose concentrations and T30-T45 for insulin concentrations, there was a significant difference between white bread + 10 g alpha-cyclodextrin and the control. EFSA concluded that this study demonstrated an effect of alpha-cyclodextrin on the reduction of post-prandial glucose responses without disproportionally increasing the post-prandial insulin responses.

In a double-blinded, randomised cross-over study, Gentilcore et al. (2011) examined the effect of alpha-cyclodextrin on the glycaemic response to an oral sucrose load in 13 subjects (median age 70 years and BMI 26.9 kg/m^2). The study also measured the effect on gastric emptying (Section 14.6.2.1). Subjects consumed either 100 g sucrose dissolved in water (control) or 100 g sucrose with 10 g alpha-cyclodextrin dissolved in water (test), with a wash-out period of at least 7 days. Three subjects did not complete the study. Blood glucose and serum insulin levels were measured at baseline and up to 300 mi (5 h) after drink consumption. No statistically significant difference was found in the iAUC for glucose and insulin responses between the control and test groups. There was also no statistically significant difference in peak blood glucose ($P = 0.88$) and serum insulin ($P = 0.22$) levels between the groups. EFSA observed that this study did not show an effect on the reduction of post-prandial glycaemic responses of alpha-cyclodextrin added to sucrose.

14.6.2.1 Proposed mechanism

With respect to the proposed mechanism, the application stated that due to the structural similarity of alpha-cyclodextrin to the helical parts of starch, alpha-cyclodextrin exerts an inhibitory effect on pancreatic amylase and may also delay gastric emptying. It was also proposed that the post-prandial reduction in glycaemic and insulinaemic responses is greater if the meal contains starch rather than sucrose as the glycaemic element.

In support of the suggested mechanisms, three studies were provided; the human study by Gentilcore et al. (2011) as well as two in vitro studies (Koukiekolo et al., 2001; Oudjeriouat et al., 2003) investigating the inhibitory effect of alpha-cyclodextrin on pancreatic amylase and barley amylase respectively.

In the study by Gentilcore et al. (2011), the amounts of the control and test drink remaining in the total, proximal and distal stomach were measured for up to 300 min (5 h) after consumption. The 50% gastric emptying time (T_{50}) was also determined. It was reported that the gastric emptying time was not significantly different between the control sucrose drink and the test drink containing sucrose + 10-g alpha-cyclodextrin.

Koukiekolo et al. (2001) describes kinetic studies carried out to understand the inhibitory effect of alpha-cyclodextrin on porcine pancreatic amylase, where amylose (one of the two polymers found in starch, the other being amylopectin) was used as a substrate. Competitive inhibition of amylose hydrolysis was demonstrated, with an inhibition constant of 7.0 mM. The inhibitory effect of alpha-cyclodextrin on barley alpha-amylase isozymes, using DP-4900 amylose as the substrate, was also examined using kinetic studies (Oudjeriouat et al., 2003). The results were compared with the effects of acarbose, known as a strong inhibitor of pancreatic alpha-amylase. Compared with acarbose, it was determined that alpha-cyclodextrin was a weak inhibitor of barley alpha-amylase isozymes.

EFSA's opinion also outlined three other studies in relation to the proposed mechanism. Similar to Oudjeriouat et al. (2003), a more recent study by Rejzek et al. (2011) reported that alpha-cyclodextrin had a weak inhibitory effect on barley β-amylase when soluble starch was used as the substrate. Two studies using crystallographic analysis are also described and in the first, Larson et al. (1994) determined that three alpha-cyclodextrin molecules bind to the alpha-amylase enzyme, two of which are in the active site. Larson et al. (2010), through crystallographic analyses of porcine pancreatic alpha-amylase with alpha-cyclodextrin, proposed a model for the binding of polysaccharides with a similar helical structure, as in natural substrates (i.e. starch and glucagon). It is further described that alpha-cyclodextrin is considered chemically identical to amylose, where its cyclic structure resembles the six glycosyl-residue turn in the amylose helix. Therefore, it is considered plausible that the association of alpha-amylase with cyclodextrins reflects the binding of helical turns of natural substrate, for example amylose.

14.6.3 Summary of evidence

On consideration of the totality of the evidence, EFSA observed that two human intervention studies (Buckley et al., 2006; Diamantis and Bär, 2002) demonstrated a significant effect of alpha-cyclodextrin added to starch on post-prandial glycaemic responses and that there was no disproportionate increase in post-prandial insulinaemic responses. Furthermore, one human study did not show an effect of alpha-cyclodextrin on post-prandial glycaemic responses when added to sucrose. EFSA also considered that the proposed mechanism, an inhibitory effect of alpha-cyclodextrin on pancreatic alpha-amylase, was in keeping with evidence provided by the three intervention studies, that is two studies carried out with starch demonstrated an effect of alpha-cyclodextrin on post-prandial glucose responses, whereas the study carried out with sucrose did not show an effect.

14.7 Conditions of use

The study by Diamantis and Bär (2002) demonstrated an attenuation of post-prandial glycaemic responses when subjects consumed 10 g alpha-cyclodextrin with 50 g starch, compared with a 50 g starch control. Furthermore, Buckley et al. (2006) examined the dose–response of post-prandial glucose levels to a 50 g starch load containing 0, 2, 5 or 10 g alpha-cyclodextrin. It was determined that the glucose response was significantly lower after consumption of both 5 g ($P = 0.03$) and 10 g ($P = 0.001$) alpha-cyclodextrin when compared with the control (0 g alpha-cyclodextrin).

In its opinion, EFSA considered that to obtain the claimed effect, at least 5 g alpha-cyclodextrin/50 g starch should be consumed. In the EU register of claims (http://ec.europa.eu/nuhclaims) these conditions remain, with the extension that 'the claim may be used for food which contains at least 5 g of alpha-cyclodextrin per 50 g of starch in a quantified portion as part of the meal'. Additionally, the conditions in the register state that 'information shall be given to the consumer that the beneficial effect is obtained by consuming alpha-cyclodextrin as part of the meal'.

14.8 Consumer issues and product development

The target population for this claim are individuals who want to reduce their post-prandial glucose levels. It is considered that this may be beneficial to those with impaired glucose tolerance (also referred to as pre-diabetes), which is common in the general population, for example the 2011 estimate of pre-diabetes for adults in England was 35% (Mainous et al., 2014). Elevated post-prandial glucose levels and pre-diabetes are associated with an increased risk of developing chronic diseases such as obesity, type 2 diabetes and cardiovascular disease. Therefore, improvements in post-prandial glycaemic responses and glucose tolerance may have long-term positive health benefits. The current World Health Organization diagnostic criteria for impaired glucose tolerance are:

- a fasting plasma glucose of less than 7.0 mmol/L and
- a 2-h venous plasma glucose (after ingestion of a 75-g oral glucose load) of ≥7.8 and <11.1 mmol/L (National Institute for Health and Care Excellence, 2012).

However, pre-diabetes presents few obvious symptoms and can go undetected for many years. In addition to the adverse health implications for individuals, this may present an issue for manufacturers in targeting individuals who want to reduce their post-prandial glucose levels. Additionally, alpha-cyclodextrin is unlikely to be familiar to the general population, which could be a barrier to purchasing intentions for products bearing this claim.

Consumer understanding of alpha-cyclodextrin and its health benefits has not been studied specifically; however, Wills et al. (2012) previously reviewed research on consumers and health claims (including understanding, attitudes and purchasing behaviour). The study reported that consumers are likely to have a more favourable opinion of products carrying health claims when the consumer is familiar with the functional ingredient (e.g. calcium), when the claim is related to general function (e.g. reduced blood glucose responses) and/or when the claim is associated with a product that has a generally positive image. Additionally, Vincent van Buul and Brouns (2015) found that consumers have a more positive opinion when they need the product, accept the ingredient, understand the beneficial effect and trust the brand.

When marketing products with health claims, manufacturers must ensure that they meet the conditions of the Health Claims Regulation, including use of the approved claim wording. However, they may also be able to draw on previous research to make the product more appealing to the target consumer. For example the claim may appeal to a wider range of consumers if alpha-cyclodextrin is added to familiar products which also present a positive health image. Products bearing the claim must contain at least 5 g alpha-cyclodextrin/50 g starch; therefore, items such as pasta, rice, bread, potatoes and noodles (or meals based on these foods) may be suitable target carrier products.

In the European Union, adults are recommended to consume 45%–60% of their total energy intake as carbohydrates, with a preference towards starchy carbohydrates such as pasta and rice. Therefore, in a typical EU diet, starchy products are likely to be common and provide a wide range of opportunities for manufacturers to develop products that utilise this claim. Indeed, UK data from the National Diet and Nutrition Survey revealed that these recommendations are met, with adults aged 19–64 years receiving 49% of the daily energy intake from carbohydrates (Public Health England, 2016). Furthermore, wholegrain or high-fibre varieties of starchy carbohydrates may provide the best fit, as their fibre content often allows them to carry other health claims as well as nutrition claims. Consumers purchasing these products may already choose them for their health benefits and be used to reading health messages on pack. Additionally, alpha-cyclodextrin is a soluble fibre and will therefore boost the fibre content of products it is added to.

14.9 Conclusions

The Article 13.1 general function health claim ‘Consumption of alpha-cyclodextrin as part of a starch-containing meal contributes to the reduction of the blood glucose rise after that meal’ was one of a unique group of claims that, after initial assessment, was deemed to provide ‘insufficient evidence’ (EFSA, 2010). These claims were eligible for resubmission, provided additional evidence was presented. Following this process, the claim reviewed here was given a positive opinion (EFSA, 2012) and was awarded in June 2013 (EU, 2013). This opinion was based on five studies, three of which were human intervention studies and two of which were in vitro studies.

Impaired glucose tolerance (pre-diabetes) is common in the general adult population and is linked to an increased risk of developing chronic diseases such as type 2 diabetes, cardiovascular disease and obesity. It is considered by EFSA that the reduction of post-prandial glycaemic responses may be beneficial to individuals with impaired glucose tolerance.

However, this claim may present some challenges for food manufacturers. Firstly, raised blood glucose responses can remain undetected for some time and the target consumers may not be aware of their needs to reduce their post-prandial glycaemic responses. Secondly, the functional ingredient alpha-cyclodextrin is unlikely to be familiar to the general consumer. However, previous research indicates that familiar products, particularly those associated with positive health benefits, are good carriers for health claims. To bear this claim, products must contain 5 g alpha-cyclodextrin/50 g starch. Therefore, wholegrain starchy foods, which often carry nutrition and health claims related to fibre, may be a suitable carrier product for this claim.

References

Binns, N., 2014. The regulation of health claims in Europe. Sadler, M.J. (Ed.), In: Foods, Nutrients and Food Ingredients With Authorised EU Health Claims, vol. 1, Woodhead Publishing, Cambridge, United Kingdom.

Buckley, J.D., Thorp, A.A., Murphy, K.J., Howe, P.R., 2006. Dose-dependent inhibition of the post-prandial glycaemic response to a standard carbohydrate meal following incorporation of alpha-cyclodextrin. Ann. Nutr. Metab. 50, 108–114.

Comerford, K.B., Artiss, J.D., Jen, K.L., Karakas, S.E., 2011. The beneficial effects of alpha-cyclodextrin on blood lipids and weight loss in healthy humans. Obesity 19, 1200–1204.

Diamantis, I., Bär, A., 2002. Effect of alpha-cyclodextrin on the glycemic index (GI) and insulinemic index (II) of starch in human volunteers. Unpublished study protocol.

EFSA Panel on Dietetic Products, Nutrition Allergies (NDA), 2010. Scientific Opinion on the substantiation of health claims related to alpha-cyclodextrin and reduction of post-prandial glycaemic responses (ID 856, 2926), maintenance or achievement of a normal body weight (ID 2925) pursuant to Article 13(1) of Regulation (EC) No. 1924/2006. EFSA J. 8 (2), 1479.

EFSA Panel on Dietetic Products, Nutrition Allergies (NDA), 2012. Scientific Opinion on the substantiation of health claims related to alpha-cyclodextrin and reduction of post-prandial glycaemic responses (ID 2926, further assessment) pursuant to Article 13(1) of Regulation (EC) No. 1924/2006. EFSA J. 10 (6), 2713.

European Union, 2013. Commission Regulation (EU) No. 536/2013 amending Regulation (EU) 432/2012 establishing a list of permitted health claims made on foods other than those referring to the reduction of disease risk and to children's development and health. Off. J. Eur. Union, L160/4-8.

Gentilcore, D., Vanis, L., Teng, J.C., Wishart, J.M., Buckley, J.D., Rayner, C.K., Horowitz, M., Jones, K.L., 2011. The oligosaccharide alpha-cyclodextrin has modest effects to slow gastric emptying and modify the glycaemic response to sucrose in healthy older adults. Br. J. Nutr. 106, 583–587.

Grunberger, G., Jen, K.L., Artiss, J.D., 2007. The benefits of early intervention in obese diabetic patients with FBCx: a new dietary fibre. Diabetes/Metab. Res. Rev. 23, 56–62.

Koukiekolo, R., Desseaux, V., Moreau, Y., Marchis-Mouren, G., Santimone, M., 2001. Mechanism of porcine pancreatic alpha-amylase. Inhibition of amylose and maltopentaose hydrolysis by alpha-, beta- and gamma-cyclodextrins. Eur. J. Biochem. 268, 841–848.

Larson, S.B., Greenwood, A., Cascio, D., Day, J., McPherson, A., 1994. Refined molecular structure of pig pancreatic alpha-amylase at 2.1 A resolution. J.Mol. Biol. 235, 1560–1584.

Larson, S.B., Day, J.S., McPherson, A., 2010. X-ray crystallographic analyses of pig pancreatic alpha-amylase with limit dextrin, oligosaccharide, and alpha-cyclodextrin. Biochemistry 49, 3101–3115.

Li, S., Shin, H.J., Ding, E.L., van Dam, R.M., 2009. Adiponectin levels and risk of type 2 diabetes: a systematic review and meta-analysis. JAMA 302, 179–188.

Mainous, A.G., Tanner, R.J., Baker, R., Zayas, C.E., Harle, C.A., 2014. Prevalence of prediabetes in England from 2003 to 2011: population-based, cross-sectional study. Br. Med. J. Open 4, e005002.

National Institute for Health and Care Excellence, 2012. National Institute for Health and Care Excellence (NICE) (2012). Type 2 diabetes: prevention in people at high risk. Public health guideline [PH38]. National Institute for Health and Care Excellence, London.

Ochiai, H., Ooka, H., Shida, C., Ishikawa, T., Inoue, D., Okazaki, R., 2008. Acarbose treatment increases serum total adiponectin levels in patients with type 2 diabetes. Endocr. J. 55, 549–556.

Oudjeriouat, N., Moreau, Y., Santimone, M., Svensson, B., Marchis-Mouren, G., Desseaux, V., 2003. On the mechanism of alpha-amylase. Eur. J.Biochem. 270, 3871–3879.

Public Health England, 2016. National Diet and Nutrition Survey. Results from Years 5 and 6 (combined) of the Rolling Programme (2012/2013-2013/2014). PHE Publications, United Kingdom.

Rejzek, M., Stevenson, C.E., Southard, A.M., Stanley, D., Denyer, K., Smith, A.M., Naldrett, M.J., Lawson, D.M., Field, R.A., 2011. Chemical genetics and cereal starch metabolism: structural basis of the non-covalent and covalent inhibition of barley beta-amylase. Mol. Biosyst. 7, 718–730.

Sadler, M.J., 2015. Authorised EU health claim for dried plums/prunes. Sadler, M.J. (Ed.), Foods, Nutrients and Food Ingredients with Authorised EU Health Claims, vol. 2, Woodhead Publishing, Cambridge, United Kingdom.

Shimazu, T., Inami, N., Satoh, D., Kajiura, T., Yamada, K., Iwasaka, T., Nomura, S., 2009. Effect of acarbose on platelet-derived microparticles, soluble selectins, and adiponectin in diabetic patients. J.Thromb.Thrombol. 28, 429–435.

Venn, B.J., Green, T.J., 2007. Glycemic index and glycemic load: measurement issues and their effect on diet-disease relationships. Eur. J. Clin. Nutr. 61 (Suppl. 1), S122–S131.

Vincent van Buul, V., Brouns, F., 2015. Nutrition and health claims as marketing tools. Crit. Rev. Food Sci. Nutr. 55 (11), 1552–1560.

Weyer, C., Funahashi, T., Tanaka, S., Hotta, K., Matsuzawa, Y., Pratley, R.E., Tataranni, P.A., 2001. Hypoadiponectinemia in obesity and type 2 diabetes: close association with insulin resistance and hyperinsulinemia. J. Clin. Endocrinol. Metab. 86, 1930–1935.

Wills, J., Storcksdieck genannt Bonsmann, S., Kolka, M., Grunert, K., 2012. European consumers and health claims: attitudes, understanding and purchasing behaviour. Proc. Nutr. Soc. 71 (2), 229–236.

Yun, J.E., Sull, J.W., Lee, H.Y., Park, E., Kim, S., Jo, J., Lee, S.J., Kim, S.Y., Choi, Y.J., Jee, S.H., Huh, K.B., 2009. Serum adiponectin as a useful marker for metabolic syndrome in type 2 diabetic patients. Diabetes/Metab. Res. Rev. 25, 259–265.

Authorised EU health claim for carbohydrates and maintenance of normal brain function

15

Ursula Arens
Freelance nutrition writer, London, United Kingdom

15.1 Introduction

The brain is almost entirely dependent on a constant supply of glucose for energy, and compared with other organs of the body, requires a disproportionately large amount. Although the weight of the brain is only about 2% of body weight, the energy requirement is about 20% of that of the whole body (Benton et al., 1996). Further, the energy stores of the brain are very small compared with its high requirements, so the availability of glucose to the brain must be uninterrupted and constant.

The supply of glucose to the brain is obtained from the glucose in the blood which in turn is obtained either from body stores of carbohydrate (glycogen in liver and muscle) and ultimately from any source of carbohydrate in the diet (starches/sugars). The body maintains and regulates blood glucose levels via several systems, mainly through the actions of insulin and glycogen. A certain amount of carbohydrate-containing foods must be consumed regularly to maintain normal blood glucose levels and to supply the brain with sufficient glucose. Inadequate amounts may lead to adverse effects including a decline in normal brain functions, demonstrated by impaired outcomes in tests of memory and cognition (Fairclough and Houston, 2004).

There is some variation of glucose utilisation by the brain in relation to age and development. Positron emission tomography can be used to demonstrate that there is an increase in rates of glucose consumption in the brain from birth until about 4 years of age, at which time the child's cerebral cortex uses over twice as much glucose as that of an adult (Chugani, 1998). From the age of 4–10 years, these very high rates of glucose consumption are maintained. Then there is a gradual decline to reach lower adult rates by the age of 16–18 years. The suggestion is that glucose consumption by sections of the brain may match the 'critical periods' of maximal learning capacity. Another suggestion from research is that some declines in cognitive performance in old age may relate to changes to glucose metabolism and regulation. Tsiakas (2007) reported that subjects over the age of 55 years with worse glucose regulation, performed worse on measures of working memory, executive function and processing speed, compared with younger

Foods, Nutrients and Food Ingredients with Authorised EU Health Claims. http://dx.doi.org/10.1016/B978-0-08-100922-2.00015-2
Copyright © 2018 Elsevier Ltd. All rights reserved.

subjects or those with better glucose regulation. Tsiakas (2007) also concluded that cognitive functioning might be compromised before gluco-regulatory impairments reach the levels leading to the diagnosis of type 2 diabetes. Conversely, there are suggestions that cognitive performance in healthy elderly persons can be enhanced with timed carbohydrate consumption (Benton, 2001; Das, 2001; Kaplan et al., 2000).

The Dietetic Products, Nutrition and Allergies (NDA) Panel of the European Food Safety Authority (EFSA) reviewed the submitted data for a health claim for carbohydrates to be made on this basis and awarded a positive scientific opinion. However, there are some concerns about the inadvertent promotion of foods that contain high amounts of sugars, and there has been a discussion of wording to more clearly capture the association between carbohydrates and brain function.

15.2 Article 13.1 claim on carbohydrates and maintenance of normal brain function

Article 13.1 claims refer to the role of a nutrient in growth, development and/or a function in the body. For this claim, the role of the nutrient carbohydrate was assessed in relation to the health effect of normal brain function. The EFSA NDA Panel adopted its opinion on health claims related to glycaemic carbohydrates and maintenance of normal brain function in April 2011 (EFSA, 2011).

15.2.1 Characterisation of glycaemic carbohydrates

The NDA Panel described the classification of carbohydrates according to their degree of polymerisation. These can be simple (mono- and disaccharides) or complex (oligo- and polysaccharides). The main glycaemic carbohydrates in the diet are glucose and fructose (monosaccharides), sucrose and lactose (disaccharides), malto-oligosaccharides and starch (polysaccharides). After consumption and digestion, all glycaemic (blood glucose-forming) carbohydrates are absorbed via the small intestine into the blood circulation and then distributed to all organs of the body to supply energy. Glucose delivered via the blood also supplies the brain with its' energy supply. The NDA Panel concluded that glycaemic carbohydrates were sufficiently characterised in relation to the claimed effect.

15.2.2 Claimed health effects for carbohydrates

The proposed claimed effects were:

- mental performance
- cognitive/mental performance
- alertness
- attention
- memory.

The NDA Panel assumed that the claimed effects refer to the maintenance of normal brain function which they considered to be a beneficial physiological effect. The Panel also assumed that the target population for communication of the proposed claim is the general population. The Panel concluded that a cause-and-effect relationship had been established between the consumption of glycaemic carbohydrates and maintenance of normal brain function, and that this was a beneficial physiological effect.

The Panel considered recommendations of population carbohydrate requirements from EU and US dietary reports. They concluded that a daily intake of 130 g of dietary carbohydrates from a balanced diet is sufficient to cover glucose requirements of the brain for both children over the age of 1 year and adults. Such amounts can be consumed as part of a balanced diet in the general population.

In conclusion, the Panel proposed the following wording to reflect the scientific evidence: 'Carbohydrates contribute to the maintenance of normal brain function', and this wording has been authorised for use in the European Union.

15.3 Commission Regulation (EU) No. 1018/2013 of 23 October 2013

After the favourable assessment of the Article 13.1 claim on the effects of carbohydrate on the maintenance of brain function, a number of Member States raised concerns that such communications would promote and encourage the consumption of foods containing sugars other than those naturally occurring. Furthermore, it would convey a conflicting and confusing message to consumers, especially in relation to general dietary advice to reduce sugar consumption.

It was agreed that some additional conditions of use of the claim and an additional statement or warning is required in relation to this specific health claim (EU Commission, 2013). Also, the claim would be limited to foods that are low in sugars or in which sugars have not been added (although they may contain sugars naturally). To bear the claim for carbohydrates and brain function, the conditions of use would be as follows:

- Information should be given to the consumer that the beneficial effect is obtained with a daily intake of 130 g of carbohydrates from all sources.
- The food must contain at least 20 g of carbohydrates metabolised by humans (excluding polyols) per quantified portion. This amount (20 g) was agreed as a result of being about 15% of the total amount of 130 g. The 15% cut-off was analogous to the declaration of vitamins and minerals on foods, suggesting 'significant' amounts.
- The food must comply with the nutrition claim Low Sugars (less than 5 g/100 g for solids or 2.5 g/100 mL for liquids) or with No Added Sugars (mono- or disaccharides, or any other food used for sweetening properties).
- The claim cannot be used on foods containing 100% sugars.

The claim was authorised in 2013 (EU Commission, 2013), and the Annex to Regulation 423/2012 was amended to include the claim.

15.4 Other claimed health effects for glucose

The European Commission had also been asked to authorise a number of health claims specifically for glucose, other than for cognitive function. These were submitted in 2011 under Article 13.5 as health claims based on newly developed scientific evidence by a company producing glucose tablets (Dextro Energy GmhH & Co. KG). The glucose claims that had been given a positive opinion by the NDA Panel of EFSA (EFSA, 2012) were:

- Metabolised within body's normal energy metabolism.
- Supports normal physical activity.
- Contributes to normal energy-yielding metabolism.
- Contributes to normal energy-yielding metabolism during exercise.
- Contributes to normal muscle function.

However, in January 2015, the Commission published a Regulation refusing permission for all of these claims (EU Commission, 2015). This was because the claims were considered to be ambiguous and misleading, as they conflict with public health advice to reduce sugar consumption. The company Dextro Energy GmbH challenged this decision, but in March 2016, the European General Court upheld the European Commission decision banning the claims, concluding that the Commission did not err in finding the claims incompatible with generally accepted principles of nutrition and health. The company Dextro Energy GmbH then submitted an appeal to the EU Court of Justice for a final consideration of the proposed glucose claims (Nutraingredients, 2016), the result of which was to uphold the opinion of the European General Court.

15.5 Article 13.5 claim on carbohydrates and contribution to normal cognitive function

Following authorisation of the Article 13.1 claim for maintenance of normal brain function, the EFSA NDA Panel was also asked to deliver an opinion on the scientific substantiation of a health claim related to glycaemic carbohydrates and contribution to normal cognitive function. This was prompted by a further application from Dextro Energy GmhH & Co. KG submitted via the Competent Authority of Germany. The scope of the application was proposed to fall under health claims based on newly developed scientific evidence (Article 13.5 claims). The first application to the NDA Panel was received in August 2014, and EFSA adopted a positive scientific opinion on the proposed claim in February 2015 (EFSA, 2015).

15.5.1 Characterisation of the food/constituent

The applicant proposed glucose as the subject of the health claim. EFSA interpreted this as glycaemic carbohydrates, as these are the dietary sources of glucose (EFSA, 2015). EFSA considered these to be sufficiently characterised in relation to the claim. Further aspects described are identical to descriptions considered in the Article 13.1 claim for carbohydrates.

15.5.2 *Claimed health effects and proposed conditions of use*

The Panel concluded that contribution to normal cognitive function is a beneficial physiological effect, encompassing several domains such as alertness, attention, memory, intelligence, learning, problem solving and language. The Panel concluded that a cause-and-effect relationship was established for the claim and the proposed wording was 'glycaemic carbohydrates contribute to normal cognitive function' (EFSA, 2015).

The conditions of use proposed by the Panel were a daily intake of 130 g of glycaemic carbohydrates, as for the authorised Article 13.1 claim for maintenance of normal brain function.

15.5.3 *Outcome of authorisation processes*

However, during discussions concerning authorisation of the claim, Member State Representatives and the Commission were mindful of the EU General Court case (Section 15.4) in which the court ruled that it was acceptable for the Commission to refuse authorisation of claims for glucose despite being awarded a positive opinion by EFSA. As the application was originally for glucose, it was concluded that it would be unfair to the applicant to authorise the claim for use on non-glucose carbohydrates.

15.6 Further application for maintenance of normal brain function claim

A further development is that EFSA was also asked to deliver a scientific opinion on an application for glucose, glycaemic carbohydrates and dextrose tablets (composed of 89% glucose monohydrate) and 'contributes to the maintenance of normal brain function'. The application under Article 13.5 as a health claim based on newly developed scientific evidence was also submitted by Dextro Energy GmhH & Co. KG. As glucose (dextrose) is one of the main glycaemic carbohydrates in the diet, EFSA interpreted the subject of the claim as glycaemic carbohydrates (EFSA, 2015), and concluded that this claim had already been assessed by the Panel with a favourable outcome.

15.7 Consumer issues

Consumers may be familiar with the term carbohydrate, although it may be in relation to discussions of weight control or sports performance, rather than aspects of brain function or cognition. Currently all packaged foods in Europe are labelled with nutrition information, and total carbohydrate content (as well as the component total sugars) is part of the standard presentation. So if consumers are interested in the carbohydrate content of many of the foods they consume (those that have packaging), information is clearly presented and easy to find. Carbohydrates are found in a wide variety of foods consumed (Table 15.1).

Table 15.1 Ten grams of carbohydrate amounts in foods

Sources	Weights (g)	Amounts
White bread	20	1 Small thin slice
Boiled potato	60	1 Hen-egg size portion
Boiled rice	30	1 Level tablespoon
Boiled pasta	50	1 Tablespoon
Apple or pear	100	1 Medium
Orange	120	1 Small
Banana	50	Half small banana
Grapes	70	15 Large
Fruit juice	100	Half glass
Milk	200	1 Glass

Typical diets of most people contain adequate amounts of carbohydrate, as this is the main energy-providing nutrient in most diets, in the forms of both sugars and starches. Many national dietary intake surveys have measured intakes of carbohydrates. In a national dietary intake survey of German adults, median intakes of total carbohydrates in men aged 19–65 years ranged from 236 to 260 g, and in women, the range was from 195 to 214 g (Max Rubner Institut, 2008). In UK adults (aged 19–64 years) median daily intakes of total carbohydrates were reported as 243 g in men and 193 g in women (Bates et al, 2014). However, 2.5 of UK adults have total carbohydrate intakes of 101 g or less (men, 122 g; women, 92 g), and hence have levels that are lower than the guidance amounts in the carbohydrate health claim, which is 130 g/day.

There may be situations in which some consumers notice temporary or occasional declines in cognitive function. If alertness or memory impairments are more frequent, or become noticeable to others, then medical and psychological assessments are urgently needed. While some nutrient deficiencies may result in cognitive declines (e.g. folate and vitamin B_{12}), these are unlikely to occur until nutrient status is impaired, and are usually associated with inadequate unbalanced diets or with impairments in nutrient absorption, for example lower rates of vitamin B_{12} absorption that may occur in old age. If this is a concern, then a dietary assessment by a nutritional professional such as a dietitian is helpful and so is the guidance on the possible use of nutrient supplements.

Less-efficient cognitive functioning due to lower levels of blood glucose may occur during situations where intakes of carbohydrate foods are inadequate to match energy demands. This may occur due to self-selected restraint of intakes such as during fasting, following very low carbohydrate diets for weight loss, or diabetes control. Alternatively, it may arise from unintended low intakes of carbohydrate due to non-availability of foods/meals or illness, or when energy demands are high such as those that occur during sports or extreme manual labour. Sustained abnormal low blood glucose may lead to hypoglycaemia which in turn leads to confusion and drowsiness and eventually coma. However, this is a specific medical situation due to impaired glucose metabolism which may occur during undiagnosed or poorly controlled diabetes or due to excessive insulin or hypoglycaemic medication, and will not occur normally.

Most people frequently consume adequate diets providing sufficient amounts of carbohydrate to exceed the level of 130 g suggested in the claim on carbohydrates and the maintenance of normal brain function. Where blood glucose levels are within the normal range, further intakes of carbohydrates are not likely to enhance cognitive functioning. However, where there is a decline in blood glucose levels (hypoglycaemia), there may be impairments to cognitive function that may be countered with the intake of a carbohydrate source.

There are concerns that the diets of many people contain too much energy (kilocalories) and that population overweight is related to many consequences resulting in premature morbidity and mortality. There are particular concerns that diets which are high in free sugars are associated with overconsumption of energy, and some population dietary guidelines advise limits to intakes.

A report on carbohydrates issued by the UK Scientific Advisory committee on Nutrition (SACN, 2015) recommends that about 50% of total dietary energy intakes should come from carbohydrates but that, of this, energy intakes from free sugars should be 5% or less. Intakes of total carbohydrates in UK adults are about 46% of dietary energy, slightly lower than the SACN recommendations, and intakes of free sugars are just over 11%, so about double the SACN recommendations. Public health advice on diet to UK consumers will be to maintain or slightly increase energy intakes from starchy foods and to significantly reduce energy intakes from foods containing free sugars, particularly sugar-sweetened drinks.

In 2010, EFSA published a scientific opinion on Dietary Reference Values for carbohydrates, for the European Union, and proposed 45%–60% of energy as the reference intake for total carbohydrates for adults and children over 1 year of age. However, it was concluded that there were insufficient data to set an upper limit for (added) sugars intake (EFSA, 2010).

15.8 Conclusions

Carbohydrates are essential dietary components, and in the diets of most people, form the major source of food energy. The brain is dependent on glucose for its' energy supply, and blood glucose is the constant source. Glucose levels in blood come ultimately from carbohydrates consumed in the diet, but the body is able to maintain and regulate blood glucose levels via several systems, mainly by the production of insulin and glycogen. In this way, blood glucose levels are kept reasonably constant so that adequate supplies for physiological functions including brain and cognitive function can be maintained.

Health claims linking carbohydrates with normal brain function have been given a positive scientific assessment by the NDA panel of EFSA and authorised for use by the EU Commission, and may be used on the labels of some carbohydrate containing foods, providing certain conditions of use are met. The claims support general public health advice that carbohydrate foods are an essential part of the balanced diet, and should be consumed regularly (with preference for greater intakes of wholegrain and high-fibre sources of starch, and lesser intakes of foods containing free sugars).

References

Bates, et al., 2014. National Diet and Nutrition Survey. Headline Results From Years 1,2, 3 And 4 of the Rolling Programme (2008/9-2011/12). Public Health England, United Kingdom.

Benton, D, 2001. The impact of the supply of glucose to the brain on mood and memory. Nutr. Rev. 59 (1), S21–S22.

Benton, D., Parker, P.Y., Donohue, 1996. The supply of glucose to the brain and cognitive functioning. J. Biosoc. Sci. 28 (4), 463–479.

Chugani, H.T., 1998. A Critical period of brain development: studies of cerebral glucose utilization with PET. Prev. Med. 27 (2), 184–188.

Das, U.N., 2001. Cognitive performance and glucose. Am. J. Clin. Nutr. 74 (3), 409.

EFSA, 2010. Scientific Opinion on Dietary Reference Values for carbohydrates and dietary fibre. EFSA J. 8 (3), 1462.

EFSA, 2011. Scientific Opinion on the substantiation of health claims related to glycaemic carbohydrates and maintenance of normal brain function (ID603, 653) pursuant to Article 13 (1) of Regulation (EC) No. 1924/2006. EFSA J. 9 (6), 2226, (13 p.).

EFSA, 2012. Scientific Opinion on the substantiation of health claims related to glucose and contribution to energy-yielding metabolism pursuant to Article 13(5) of Regulation (EC) No. 1924/2006. EFSA J. 10 (5), 2694.

EFSA, 2015. Scientific Opinion on the substantiation of a health claim related to glycaemic carbohydrates and contribution to normal cognitive function pursuant to Article 13(5) of Regulation (EC) No. 1924/2006. EFSA J. 13 (2), 4026–4027, (7 p.).

EU Commission, 2013. Commission Regulation (EU) No. 1018/2013 of 23 October 2013 amending Regulation (EU) No. 432/2012 establishing a list of permitted health claims made on foods other than those referring to the reduction of disease risk and to children's development and health. Off. J. Eur. Union L282, 43–45.

EU Commission, 2015. Commission Regulation (EU) 2015/8 of 6 January 2015 refusing to authorise certain health claims made on foods, other than those referring to the reduction of disease risk and to children's development and health. Off. J. Eur. Union L3, 6.

Fairclough, S.H., Houston, K., 2004. A metabolic measure of mental effort. Biol. Psychol. 66 (2), 177–190.

Kaplan, R.J., Greenwood, C.E., Wincour, G., Wolever, T., 2000. Cognitive performance is associated with glucose regulation in healthy elderly persons and can be enhanced with glucose and dietary carbohydrates. Am. J. Clin. Nutr. 72 (3), 825–836.

Max Rubner Institut (MRI), 2008. Nationale Verzehrsstudie 11, Ergebnisbericht, Teil 1. Available from: www.mri.bund.de/fileadmin/Institute/EV/NVS11/Abschlussbericht.

Nutraingredients (Starling S), 2016. Dextro Energy glucose claims contest goes into extra time at ECJ. Available from: www.nutraingredients.com/Regulation-Policy.

SACN, 2015. Scientific Advisory Committee on Nutrition, Carbohydrates and Health Report. Public Health England, United Kingdom.

Tsiakas, M., 2007. The Effect of Glucose Regulation on Cognitive Function in a Healthy Sample of Older Adults. Thesis. University of Ottawa, Ottawa. Available from: www.ruor.uottawa.ca/handle/10393/29700.

Authorised EU health claims for activated charcoal, lactulose and melatonin

16

Michele J. Sadler
Rank Nutrition Ltd, Bethersden, Kent, United Kingdom

16.1 Introduction

This chapter focuses on four health claims that have been authorised for three substances that would most likely be presented in the format of a dietary supplement, rather than being present in or added to conventional foods (Table 16.1). The claims are somewhat controversial, as the substances may be considered medicinal in at least some EU Member States.

The authorised claims for these substances highlight the situation referred to in 'Whereas 17' of Regulation EU 432/2012, establishing the list of permitted Article 13.1 'general function' health claims made on foods (EU, 2012). This intimates that the classification of products as foodstuffs or as medicinal products is governed by specific union and national legislation and the inclusion of substances in the list of permitted claims does not constitute authorisation for the marketing of the substance, any decision on whether the substance can be used in foodstuffs or its classification as a foodstuff. Each Member State thus applies its own national rules about medicinal status, such that a product that is determined to be a medicine in one country may be classified under a different category in another Member State.

Additionally, in some Member States such as the United Kingdom, whether a product is deemed medicinal or not may be determined on a case-by-case basis depending on the particular product in question, rather than by a blanket rule that the substance is always considered medicinal. This judgement depends on factors such as how the product is presented and promoted, the formulation and dose and how the product may be perceived by the consumer.

In Member States where substances are viewed as medicinal, or have the potential to be viewed as medicinal products, it is unlikely they would be presented under food law or be eligible to bear the claims. This underlines the position that it is the role of European Food Safety Authority (EFSA) to assess the scientific evidence presented, and not to judge whether a health claim can be authorised or not, or even if it is authorised whether the substance is subject to other overriding legislation which might mean in practice that the claim cannot be used.

Foods, Nutrients and Food Ingredients with Authorised EU Health Claims. http://dx.doi.org/10.1016/B978-0-08-100922-2.00016-4
Copyright © 2018 Elsevier Ltd. All rights reserved.

Table 16.1 **Summary of authorised claim wording and conditions of use**

Substances	Authorised claims wording	Conditions of use
Activated charcoal	Activated charcoal contributes to reducing excessive flatulence after eating	The claim may be used only for food that contains 1 g of activated charcoal/quantified portion. To bear the claim, information shall be given to the consumer that the beneficial effect is obtained with 1 g which should be taken at least 30 min before and 1 g shortly after the meal
Lactulose	Lactulose contributes to an acceleration of intestinal transit	The claim may be used only for food that contains 10 g of lactulose in a single quantified portion. To bear the claim, information shall be given to the consumer that the beneficial effect is obtained with a single serving of 10 g lactulose/day
Melatonin	Melatonin contributes to the alleviation of subjective feelings of jet lag	The claim may be used only for food that contains at least 0.5 mg of melatonin/quantified portion. To bear the claim, information shall be given to the consumer that the beneficial effect is obtained with a minimum intake of 0.5 mg to be taken close to bedtime on the first day of travel and on the following few days after arrival at the destination
Melatonin	Melatonin contributes to the reduction of time taken to fall asleep	The claim may be used only for food that contains at least 1 mg of melatonin/quantified portion. To bear the claim, information shall be given to the consumer that the beneficial effect is obtained by consuming 1 mg of melatonin close to bedtime

The claims covered in this chapter are for activated charcoal, lactulose (4-O-β-D-galactopyranosyl-β-D-fructofuranose) and melatonin (*N*-acetyl-5-methoxy tryptamine) and are examples of claims that have been authorised but are not in widespread use, as they have the potential to be considered medicinal in some Member States.

16.2 Activated charcoal

16.2.1 Characterisation of the substance

Activated charcoal, also referred to as activated carbon, is charcoal that has been treated with oxygen to make it more porous. This treatment results in the development of microscopic holes which significantly increases the surface area. Liquids or gases that pass through activated charcoal bind chemically to the exposed carbon by the process of adsorption, filling the active sites. Adsorption increases as pH decreases, and longer contact with activated charcoal also increases its effectiveness. Activated charcoal is a fine black powder consisting of pure carbon that is both tasteless and odourless. It is usually retailed as such or in an encapsulated format.

Activated charcoal was discovered in ancient times and used to manage a range of ailments. Today it has a number of medical uses including the treatment of orally ingested poisons and overdoses of drugs. It prevents their absorption by binding to the harmful compounds which are then excreted in the stools. In medicines and over-the-counter products, it is used to filter water to remove chlorine, volatile compounds causing odours such as hydrogen sulphide and some minerals such as mercury and iron.

As it is also highly effective in adsorbing gases, it has a use in controlling flatulence which is the subject of the authorised health claim.

16.2.2 EFSA scientific opinion

As a part of the assessment of potential health claims under Article 13.1 (EC, 2006), the EFSA panel on Dietetic Products, Nutrition and Allergies (NDA) was asked to give an opinion on a possible health claim for activated charcoal and 'gastrointestinal health'. Following EFSA's initial screening of submitted claims to assess their suitability for review (Binns, 2014), this claim was returned for further clarification. Further to this process and with additional information received, the NDA panel interpreted the health relationship as two possible claims: (1) 'reduction of excessive gas accumulation' and (2) 'reduction of bloating'.

For the first claim, the NDA panel commented that excessive gas accumulation may cause abdominal pain and discomfort and thus concluded that reduction of excessive gas accumulation is a beneficial physiological effect. They also considered reduction of bloating to be a beneficial physiological effect. The target group for both of the claims was assumed to be the general population.

16.2.2.1 Summary of substantiating evidence

For the claim 'reduction of excessive gas accumulation', the NDA panel judged that three intervention studies published in the 1980s provided suitable evidence to substantiate the claim (Hall et al., 1981; Jain et al., 1986a,b). The studies tested multiple intakes of capsules containing 194 or 260 mg activated charcoal/capsule.

Following a gas-producing meal, Hall et al. (1981) compared the effectiveness of activated charcoal against placebo on reducing intestinal gas in two different studies. Both were double-blind, placebo-controlled trials conducted in healthy adults aged

18–40 years. The volunteers consumed either a bean-containing meal or a normal meal containing no known gas-forming constituents. The outcome measure in the first study was the number of flatus events. Baseline data in 13 volunteers were recorded after consuming a normal meal. The volunteers then consumed a bean-containing meal on 2 different days, with 2 days in between. They either took 3 capsules (194 mg activated charcoal/capsule) after the meal and 3 more capsules 2 h after the meal, or they took identical capsules containing placebo at the same time points. Additionally, seven volunteers took 2 capsules 2 h after the meal only. Subjects recorded the number of flatus events over the following 7 h and did not eat lunch or snacks over this time period. In the second study the outcome measure was breath hydrogen. Ten subjects ate either a normal meal or a bean-containing meal. Those given the bean-containing meal consumed either 3 capsules of activated charcoal or placebo immediately after the meal and 3 capsules every 30 min over the next 2 h (15 capsules overall) during which time breath hydrogen was measured. The results demonstrated that activated charcoal was effective in curtailing the increase in the number of flatus events ($P < 0.01$) and the increase in breath hydrogen concentrations ($P < 0.001$) following the bean-containing meal compared with placebo. The benefit in the subjects who only took activated charcoal 2 h after the meal was reduced in comparison to subjects who took activated charcoal immediately after the meal.

Jain et al. (1986a) investigated the effects of activated charcoal compared with placebo on reducing intestinal gas in two study populations with different dietary habits and ecology of gut flora, one in the United States and one in India. In an acute cross-over study, lactulose was used to stimulate gas production and the outcome measure was breath hydrogen. Following a 10-h fast, a baseline breath sample was collected and the subjects consumed either 4 activated charcoal capsules containing 260 mg/capsule, or placebo capsules. They ingested 15 mL lactulose 30 min after taking the capsules, followed by a second intake of capsules 30 min later. The study was repeated 1 week later, in which subjects were switched from control to test group and vice versa. Subjects who did not show a rise in breath hydrogen following lactose ingestion were excluded from the analysis (2 in the United States out of 30 subjects, and 7 in India out of 69 subjects). Compared with placebo, activated charcoal significantly reduced breath hydrogen in subjects both in the United States and in India ($P < 0.05$).

In a further double-blind, randomised trial (Jain et al., 1986b), 20 mL of end-expiratory fasting breath samples were collected from each subject to provide a baseline measure. Four capsules containing either placebo or 260 mg activated charcoal/capsule were ingested by the volunteers followed by a meal of baked beans 30 min later, and a further intake of 4 capsules 30 min after consuming the baked beans. Breath samples were collected every 30 min for 7 h. One subject out of a total of 10 did not produce breath hydrogen, and hence the analyses were for the remaining 9 subjects. Activated charcoal significantly reduced the peak breath hydrogen ($P = 0.002$) and the mean area under the curve breath hydrogen over 0–7 h ($P < 0.003$), at 2.5–3.0 h and 4.5–6.0 h following ingestion of the activated charcoal.

As there was also a plausible mechanism for the effect, that is adsorption of flatus generated in the large intestine, the NDA panel concluded that a cause-and-effect relationship was established between ingestion of activated charcoal and reduction of excessive gas accumulation, and a positive opinion was issued.

However, a cause-and-effect relationship for the reduction of bloating was not considered to have been demonstrated. Though two of the intervention studies reported symptom scores for bloating and gastrointestinal discomfort (Jain et al., 1986a,b), it was not clear that validated methodology had been used to score the subjective symptoms, and thus the second claim was given a negative opinion.

16.2.3 Authorisation of the claim

The claim for 'reduction of excessive gas accumulation' was authorised under Regulation EU 432/2012 and added to the EU register of claims. The authorised wording for the claim is 'Activated charcoal contributes to reducing excessive flatulence after eating'. The conditions of use are that the claim may only be used for food (which includes dietary supplement formats) containing 1 g of activated charcoal/quantified portion. Information should be given to the consumer that the beneficial effect is obtained by taking 1 g at 30 min prior to eating and an additional 1 g shortly afterwards.

16.2.4 How the claim is being used

The claim is used for limited number of dietary supplements. The dose of 1 g prior to a meal plus 1 g following a meal is relatively high, such that not all products on the market qualify for the claim. Hence the use of the claim may require reformulation of products to ensure the dose provides sufficient quantity of activated charcoal to meet the conditions of use. One example of a product making the claim requires the consumer to take 3 capsules prior to a meal and 3 capsules following the meal to reach the required dose. The authorised wording is used which in this case is a straightforward claim for consumers to understand.

However, in some Member States activated charcoal has the potential to be considered as medicinal by nature, particularly where licensed medicines are available, and in some Member States the claim itself may be considered medicinal. The claim is not generally used in these countries.

16.3 Lactulose

16.3.1 Characterisation of the substance

Lactulose is a synthetic disaccharide that is made either by the alkaline isomerisation of lactose or by enzyme-catalysed synthesis (EFSA, 2010a). Hence lactulose is not naturally present in any foods and is supplementary to the usual diet.

16.3.2 Substantiation of the claim

As part of the assessment of potential health claims under Article 13.1 (EC, 2006), EFSA was asked to give an opinion on a possible health claim for lactulose and 'transit effect'. The NDA panel interpreted this as 'reduction in intestinal transit time' which was considered to be a beneficial physiological effect, within the normal range.

A further claim proposed was 'probiotic/bifidogenic' which the panel interpreted as meaning an increase in the numbers of healthy or good bacteria. The panel held the view that the scientific evidence does not establish that increasing the number of colonic bacteria is a beneficial physiological effect, as the numbers of relative proportions of bacterial groups that are consistent with a healthy or balanced microflora have not been established. Additionally, the panel does not consider that increasing the number of any particular groups of microorganisms represents a beneficial physiological effect. However, the panel does consider that reducing the numbers of potentially pathogenic colonic bacteria might be a beneficial physiological effect, and hence the claim was assessed on this basis. The target group for both claims is the general population.

16.3.2.1 Summary of substantiating evidence

The evidence submitted for the claim 'lactulose reduces intestinal transit time' included five studies that the NDA panel was able to consider to substantiate the claim. Across the various studies lactulose was compared with placebo, lactitol or with inulin, and was administered at a dose of 10, 20 or 40 g in either liquid or crystalline form. Outcome measures included oro-caecal transit time, small bowel transit, breath hydrogen and the frequency weight, volume and water content of stools. The studies were undertaken in normal healthy human subjects and chronically constipated volunteers with the numbers per study ranging from 8 to 66 (Bass and Dennis, 1981; Geboes et al., 2003; Miller et al., 1997; Pontes et al., 1995; Wutzke et al., 1997).

All of the studies showed a consistent, statistically significant effect of lactulose on transit time. There is also a well-established mechanism by which lactulose exerts its effect on reducing transit time. Following ingestion, lactulose is degraded in the colon to lactic acid along with small amounts of formic and acetic acids by the action of beta-galactosidases from the resident bacteria in the colon. This results in increased osmotic pressure and mild acidification of the colonic content which causes an increase in stool water content, softening of the stools, and hence faster transit through the colon (Drugbank, 2009). The NDA panel gave a positive opinion for the claim that 'lactulose reduces intestinal transit time', based on a daily intake of 10 g lactulose.

For the second claim, 'reducing the numbers of potentially pathogenic colonic bacteria', three studies had the potential to substantiate the claim. However, one study had few subjects, used weak methodology for the microbiological analyses and was not blinded. The second study measured clostridia but did not provide evidence that this bacterial group is characterised as pathogenic. In the third study the bacterial groups analysed were also not supported by evidence that they are characterised as pathogenic. Hence the panel determined that no conclusions could be drawn from the studies in relation to the claimed effects, and a negative opinion was issued for the claim 'reducing the numbers of potentially pathogenic colonic bacteria'.

16.3.3 Claim authorisation

The claim for reduced transit time was authorised under Regulation EU 432/2012 and added to the EU register of claims. The authorised wording for the claim is 'lactulose contributes to an acceleration of intestinal transit'. The conditions of use are that the claim may only be used for food that contains 10 g lactulose/single quantified portion. The consumer must be informed that the beneficial effect is obtained with a single serving of 10 g of lactulose/day.

16.3.4 Use of the claim

Lactulose is indicated on the EU novel food catalogue as not novel in foods, that is it is an ingredient with acknowledged food use and therefore is not prohibited from use in a food or food supplement. However, the claim is controversial because in Europe lactulose is used as a medicine for the treatment of constipation. Unlicensed lactulose products that carry the claim may therefore be considered as being presented for a medicinal purpose and hence regarded as medicinal products. This may depend on factors such as how the product is presented, who it is targeted at, the dose provided and consumer perception, whereby if lactulose is known to treat constipation the consumer might think a lactulose food supplement or lactulose-containing food could provide the same benefit.

16.4 Melatonin

Further examples of this situation are the two authorised health claims for melatonin.

16.4.1 Characterisation of the substance

Melatonin is a hormone that is biosynthesised naturally by a number of tissues in the body, the main site being the pineal gland in the brain. Peak levels of production and release occur during the hours of darkness. The biosynthetic pathway is that L-tryptophan is converted to 5-hydroxy-L-tryptophan which is decarboxylated to produce serotonin. Only in hours of darkness is serotonin converted to *N*-acetyl serotonin which is finally converted to melatonin. Its secretion in the body is controlled by norepinephrine.

16.4.2 Role of melatonin in the body

Melatonin is involved in control of the circadian rhythm and regulation of the 'biological clock' including night–day cycles. It also has an antioxidant activity.

16.4.3 Substantiation of the claim

As part of the assessment of potential health claims under Article 13.1 (EC, 2006), EFSA was asked to give an opinion on a possible health relationship for 'sleep-wake cycle regulation' with proposed wordings such as 'helps to reduce jet lag effects', 'helps to reduce sleep onset latency' and 'contributes to improve sleep quality'. Further to clarification, the NDA panel interpreted this as two possible health claims. One was 'alleviation of subjective feelings of jet lag' which the panel concluded might be a beneficial physiological effect. The second was 'reduction of sleep onset latency (i.e. reduced time to fall asleep) and/or improvement of sleep quality' which the NDA panel also concluded might be beneficial physiological effects and for which there are validated methods of measurement. Both of these claims were assessed in one EFSA opinion (EFSA, 2010b).

Further proposed claims for melatonin 'sleep-wake cycle regulation', 'relaxation' and 'sleep patterns', that provided additional evidence for the claimed effects, were also interpreted as 'reduction of sleep onset latency' and these were assessed by the NDA panel in a separate opinion (EFSA, 2011).

The target group for the claims is the general population.

16.4.3.1 Summary of substantiating evidence

The NDA panel deemed that much of the evidence submitted to support the claims was unsuitable because the studies had been undertaken in subjects with primary or secondary sleep disorders. No evidence was submitted to establish that this was representative of the general population or that the evidence could be extrapolated to the general population.

The main evidence substantiating the claim for 'alleviation of subjective feelings of jet lag' was a Cochrane Systematic Review (Herxheimer and Petrie, 2002, reviewed in 2008). It included 11 studies in which the primary outcome was subjective feelings of jet lag. Doses tested ranged from 0.5 to 5 mg melatonin/day and these were shown to be effective in reducing or preventing subjective ratings of jet lag. The NDA panel thus awarded a positive opinion for the claim. They proposed a daily intake of 0.5–5 mg melatonin/day taken close to bedtime on the day of travel and on the following few days after arrival, and commented that melatonin appears to be safe with short-term use (i.e. 3 months or less).

For the initial assessment of the claim for 'reduction of sleep onset latency (i.e. reduced time to fall asleep) and/or improvement of sleep quality' no evidence was provided to support the claim, and the panel therefore considered a Cochrane Systematic Review and meta-analysis (Buscemi et al., 2006) that they were aware of. However, this only addressed efficacy for subjects with secondary sleep disorders accompanying sleep restriction, and did not consider the effects of melatonin on subjective ratings of daytime fatigue. Hence for this first assessment of this particular claim the panel concluded that a cause-and-effect relationship was not established and a negative opinion was issued for 'reduction of sleep onset latency' or 'improvement of sleep quality' (EFSA, 2010b).

However, a further assessment of 'reduction of sleep onset latency', for which additional evidence had been submitted, led to a positive assessment (EFSA, 2011). The relevant studies were reviewed in three meta-analyses (Brzezinski et al., 2005; Buscemi et al., 2004, 2005). These addressed the effect of melatonin on sleep onset latency using objective sleep outcome measures (such as polysomnography, actigraphy, index finger switch depression or questionnaires) in healthy subjects without insomnia

(Buscemi et al., 2004), subjects with primary sleep disorders (Buscemi et al., 2005) and in healthy volunteers with or without insomnia (Brzezinski et al., 2005).

The first of the three meta-analyses (Buscemi et al., 2004), was based on 20 included studies comparing the effect of melatonin versus placebo over periods of 1 night to 8 weeks on sleep onset latency. Doses ranged from 0.1 to 100 mg melatonin. The analysis showed a statistically significant decrease in the onset of sleep latency (weighted mean difference: −3.9 min; 95% CI: −5.3 to −2.6). The second review (Buscemi et al., 2005) included studies in volunteers with insomnia or delayed sleep-phase syndrome, and included children, adults (19–65 years) and older adults (66+ years). There was a statistically significant reduction in sleep onset latency with melatonin compared with placebo (weighted mean difference: −11.7 min; 95% CI: −18.2 to −5.2). The effective dose was in the range 1–3 mg melatonin, suggesting a consistent effect of 1 mg melatonin on sleep latency across the two meta-analyses.

The third meta-analysis (Brzezinski et al., 2005) included studies (n = 12) that tested doses from 0.1- to 40-mg melatonin/day, over periods from 1 night to 3 weeks, and showed a statistically significant reduction in sleep onset latency of −3.9 min (95% CI: −2.4 to −5.4) compared with placebo. Despite a number of limitations with the data, the NDA panel concluded that this analysis could also be used as evidence to substantiate the health claim.

The NDA panel noted that the meta-analysis they had considered in their previous assessment of the same claim (Buscemi et al., 2006) addressed subjects with sleep restrictions (after jet lag, or with induced insomnia or working shifts), and did not show an effect on sleep onset latency. However, the three additional meta-analyses considered in the second opinion addressed different subject groups and showed statistically significant benefits of melatonin. Hence the panel awarded a positive opinion for the claim following this second assessment of evidence (EFSA, 2011).

16.4.3.2 Mechanisms of action

Neither of the EFSA opinions refers to the potential mechanism of action of melatonin. While there is substantial understanding of the involvement of melatonin in circadian rhythms and the importance of its dynamic pattern of secretion, precisely how it mediates the regulation of sleep is unclear. In vitro studies suggest that melatonin acts on the central nervous system, modulates synchronisation of the biological clock, and improves the promotion of sleep by acting on the suprachiasmatic nucleus of the hypothalamus which may involve interaction with melatonin receptors (MT1 and MT2 subtypes). However, studies measuring in vivo effects in animals are difficult to extrapolate to humans (EMEA, 2007).

16.4.4 Authorisation of the claim

Both of the claims with positive EFSA opinions were authorised under Regulation EU 432/2012 and added to the EU register of claims. The authorised wording for the claims is ‘melatonin contributes to the alleviation of subjective feelings of jet lag’ and ‘melatonin contributes to the reduction of time taken to fall asleep’.

16.4.4.1 Conditions of use

The conditions of use for the alleviation of feelings of jet lag claim are that the quantified portion of food (which would include a dietary supplement format) must provide at least 0.5 mg of melatonin, and the consumer must be informed that this amount should be taken close to bedtime on the first day of travel and on the following few days after arrival.

The conditions of use for the reduction of sleep onset latency claim are that the quantified portion of food (which would include a dietary supplement format) must provide 1 mg of melatonin, and the consumer should be informed that the beneficial effect is obtained by ingesting 1 mg melatonin close to bedtime.

Melatonin is present in low levels in the diet, occurring in fruits (such as bananas, plums, grapes and cherries), grains (such as oats, rice, wheat and barley) and beverages (such as tea, coffee, wine and beer). However, for the amount needed to deliver the health benefit, the main source of melatonin would be from a synthetic source.

16.4.5 Use of the claim

Whilst the claims for melatonin have been authorised for use on food, the classification of melatonin is not harmonised across the European Union. Melatonin can be presented as a food supplement in some Member States. However, in other EU Member States such as France and the United Kingdom, melatonin is regarded as a medicinal substance and can only be obtained on prescription, and not as a food ingredient. Other EU countries such as Spain and Italy take the stance that whether a product has food or medicinal status depends on the dose. Italy, for example allows the use of up to 1 mg melatonin in food supplements. In contrast, a Dutch ruling has indicated that the status of melatonin can only be determined on a case-by-case basis depending on all the product characteristics and mode of consumption, and not on the basis of dose alone.

An additional issue is that the claims themselves may be deemed medicinal by the medicines agencies of individual Member States. In this case any product bearing the claims (irrespective of the dose) would have the potential to be regarded as a medicinal product. This reflects that claims promoting sleep may be regarded as medicinal claims, and in the United Kingdom, for example reference to the ‘adverse condition of jet lag’ particularly in terms of modifying the body clock and sleep cycle is likely to raise the question as to whether the product is medicinal (MHRA, 2016).

16.5 Conclusions

Four of the claims discussed in this chapter are authorised for use in the European Union. However, they are not in widespread use as the status of products bearing these claims might be considered medicinal when taking into account a range of relevant factors.

Activated charcoal requires a high intake to make the claims, and in some EU Member States the claim itself may be regarded as medicinal. Lactulose is available on prescription in some Member States such that consumer perception may be an issue with products presented as foods. Though melatonin can be presented as a food supplement in some Member States, in others it is regarded as a medicinal substance; in others its status will depend on the dose, and in others on the product presentation.

These claims demonstrate that authorisation in the European Union does not automatically allow use of the claims, and other relevant factors may need to be taken into account when deciding whether or not to make use of authorised health claims.

16.6 Sources of further information and advice

EFSA, 2016. General Scientific guidance for stakeholders on health claim applications. EFSA J. 14 (1), 4367.

The Commission website (http://ec.europa.eu/nuhclaims) includes the register of EU health claims.

The European Parliament and the Council of the European Union, 2006. Corrigendum to Regulation (EC) No. 1924/2006 of the European Parliament and of the Council of 20 December 22006 on nutrition and health claims made on foods. Off. J. Eur. Union, L 12/3.

Acknowledgements

I would like to acknowledge the assistance of Jenny Akers of Clear Check, Health Food Manufacturers' Association (HFMA; 1 Wolsey Road, East Molesey, Surrey KT8 9EL, United Kingdom) for commenting on the draft chapter, and for her valuable help regarding the potential for the medicinal status of the products discussed in this chapter.

References

Bass, P., Dennis, S., 1981. The laxative effects of lactulose in normal and constipated subjects. J. Clin. Gastroenterol. 3, 23–28.

Binns, N., 2014. The regulation of health claims in Europe. Sadler, M.J. (Ed.), Foods, Nutrients and Food Ingredients with Authorised EU Health Claims, vol. 1, Woodhead Publishing, Cambridge, p. 2014.

Brzezinski, A., Vangel, M.G., Wurtman, R.J., et al., 2005. Effects of exogenous melatonin on sleep: a meta-analysis. Sleep Med. Rev. 9, 41–50.

Buscemi, N., Vandermeer, B., Pandya, R., et al., 2004. Melatonin for Treatment of Sleep Disorders. Summary, Evidence Report/Technology Assessment No. 108. Prepared by the University of Alberta Evidence-Based Practice Center, under Contract No. 290-02-0023. AHRQ Publication No. 05-E002-1. Agency for Healthcare Research and Quality, Rockville.

Buscemi, N., Vandermeer, B., Hooton, N., et al., 2005. The efficacy and safety of exogenous melatonin for primary sleep disorders: a meta-analysis. J. Gen. Intern. Med. 20, 1151–1158.

Buscemi, N., Vandermeer, B., Hooton, N., et al., 2006. Efficacy and safety of exogenous melatonin for secondary sleep disorders and sleep disorders accompanying sleep restriction: meta-analysis. Br. Med. J. 33, 385–393.

Drugbank, 2009. Lactulose. Available from: www.drugbank.ca/drugs/DB00581.

EU, 2006. 30.12.2006 Regulation EC 1924/2006 on nutrition and health claims made on foods, and 18.1.2007 Corrigendum to Regulation EC 1924/2006 on nutrition and health claims made on foods. Off. J. Eur. Union.

European Food Safety Authority, 2010a. Scientific opinion on the substantiation of health claims related to lactulose and decreasing potentially pathogenic gastro-intestinal micro-organisms and reduction in intestinal transit time pursuant to Article 13(1) of Regulation (EC) No 1924/20061. EFSA J. 8 (10), 1806.

European Food Safety Authority, 2010b. Scientific opinion on the substantiation of health claims related to melatonin and alleviation of subjective feelings of jet lag and reduction of sleep onset latency, and improvement of sleep quality. EFSA J. 8 (2), 1467.

European Food Safety Authority, 2011. Scientific opinion on the substantiation of a health claim related to melatonin and reduction of sleep onset latency. EFSA J. 9 (6), 2241.

European Medicines Agency (EMEA), 2007. Assessment report for Circadin (melatonin). EMEA/H/C/695. Available from: http://www.ema.europa.eu/docs/en_GB/document_ library/EPAR_-_Scientific_Discussion/human/000695/WC500026808.pdf.

European Union (EU), 2012. Commission Regulation (EU) No. 432/2012 of 16 May 2012, establishing a list of permitted health claims made on foods, other than those referring to the reduction of disease risk and to children's development and health. Off. J. Eur. Union L 136/1.

Geboes, K.P., Luypaerts, A., Rutgeerts, P., Verbeke, K., 2003. Inulin is an ideal substrate for a hydrogen breath test to measure the orocaecal transit time. Aliment. Pharmacol. Ther. 18, 721–729.

Hall, Jr., R.G., Thompson, H., Strother, A., 1981. Effects of orally administered activated charcoal on intestinal gas. Am. J. Gastroenterol. 75, 192–196.

Herxheimer, A., Petrie, K.J., 2002. Melatonin for the prevention and treatment of jet lag. Cochrane Database of Syst. Rev. 2, CD001520.

Jain, N.K., Patel, V.P., Pitchumoni, C.S., 1986a. Efficacy of activated charcoal in reducing intestinal gas: a double-blind clinical trial. Am. J. Gastroenterol. 81, 532–535.

Jain, N.K., Patel, V.P., Pitchumoni, S., 1986b. Activated charcoal, simethicone, and intestinal gas: a double-blind study. Ann. Intern. Med. 105, 61–62.

Medicines and Healthcare Products Regulatory Agency (MHRA), 2016. A guide to what is a medicinal product. MHRA Guidance Note 8. Her Majesty's Stationery Office, London.

Miller, M.A., Parkman, H.P., Urbain, J.L.C., et al., 1997. Comparison of scintigraphy and lactulose breath hydrogen test for assessment of orocecal transit—lactulose accelerates small bowel transit. Digest. Dis. Sci. 42, 10–18.

Pontes, F.A., Silva, A.T., Cruz, A.C., 1995. Colonic transit times and the effect of lactulose or lactitol in hospitalized-patients. Eur. J. Gastroenterol. Hepatol. 7, 441–446.

Wutzke, K.D., Heine, W.E., Plath, C., et al., 1997. Evaluation of oro-cecal transit time: a comparison of the lactose-[^{13}C, ^{15}N]ureide $^{13}(CO_2)$- and the lactulose H_2-breath test in humans. Eur. J. Clin. Nutr. 51, 11–19.

Further Reading

European Food Safety Authority, 2011. Scientific Opinion on the substantiation of health claims related to activated charcoal and reduction of excessive intestinal gas accumulation and reduction of bloating pursuant to Article 13(1) of Regulation (EC) No 1924/20061. EFSA J. 9 (4), 2049.

Index

S

T

CPI Antony Rowe
Chippenham, UK
2017-11-03 21:17